A MANUAL OF RESINS
FOR SURFACE COATINGS

VOLUME II

Edited by **P. K. T. Oldring** BA, Ph.D
and **G. Hayward** C.Chem, MRSC

With contributions by a team of Senior Practising Chemists
within the Synthetic Resins, Paint, Printing Inks
and Coatings Industries in the UK

including

Roger Castle B.Sc
C. Standen B.Sc
G. Hayward C.Chem, MRSC

© 1987 Selective Industrial Training Associates Limited
London, United Kingdom

FIRST IMPRESSION 1987
SECOND IMPRESSION 1990
THIRD IMPRESSION 1993

Typesetting and Artwork produced by K-Dee Typesetters

Published by
SITA Technology Ltd
Gardiner House
Broomhill Road
LONDON SW18 4JQ
ENGLAND

ISBN 0.947798.06.4

INTRODUCTORY NOTE

This text book has been specifically designed for paint and ink technologists dealing with specific product related technical areas. Each aspect of the technology has been written by experts in that field. This has been done in order to provide comprehensive practical theory. In some cases this naturally involves repetition of theoretical aspects of the chemistry required for an in depth understanding of the particular technology under discussion.

A MANUAL FOR RESINS
FOR SURFACE COATINGS

CONTENTS

Chapter V

Epoxy Resins

Chapter V

Epoxy Resins

INTRODUCTION

Epoxy resins are polymers containing an epoxide group, viz:

$$\begin{array}{c} O \\ \diagup \,\, \diagdown \\ -\,CH\,-\!-\,CH\,- \end{array}$$

Epoxy resins are comparatively expensive and are only used in coating formulations when their superior properties are required.

These properties include:

> Good chemical resistance, particularly to alkaline environments
>
> Outstanding adhesion to a variety of substrates
>
> Excellent toughness, hardness and flexibility
>
> Outstanding water resistance.

Epoxy resins may be considered as reactive intermediates. Although solutions of very high molecular weight epoxy resins are used on their own, in some specialist applications, most epoxy resins require curing (cross linking into a three dimensional network) to form useful coatings. Cross linking can occur via the epoxide or hydroxyl groups of the resin. In the presence of a suitable catalyst, epoxy resins can be made to link with each other or with other polymers or high molecular weight species. It is the capability of the epoxy resin to react with a variety of reactants that gives these resins great versatility in coating applications.

Most epoxy resins used in surface coating systems have epoxide equivalent weights (E.E.W.) between 180 and 3200. Epoxy resins with E.E.W.'s of 180−475 are used mainly in 'two-package' low temperature cure systems.

Epoxy resins with E.E.W.'s in the range 700−1,000 are normally used in epoxy-ester systems.

Epoxy resins with E.E.W.'s in the range 1,500−3,200 are used in stoving finishes.

A more recent and interesting development is the use of epoxy resins with molecular weights of 50,000 or more. When these resins are applied as solutions, they have large enough molecular size to form thermoplastic films, when the solvent is evaporated.

Although films formed in this way are not cross linked (and hence do not exhibit the excellent properties normally associated with epoxy resins) they show good adhesion and chemical resistance coupled with extreme flexibility.

THE MANUFACTURE OF EPOXY RESINS

Epoxy resins are produced by the base induced reaction of epichlorhydrin with a poly-hydroxy compound, most usually diphenylol propane (Bisphenol A) (see page 5).

As the mole ratio of Bisphenol A to epichlorhydrin is increased, higher molecular weight polymers are formed. For every repeat unit **n** a pendant OH group is formed and although the epoxy resin is normally represented as a linear molecule, some branching will occur, sited on the OH groups. This is particularly true at high values of **n** (see page 6).

COMMERCIAL EPOXY RESINS

By far the most commercially significant epoxy resins are those obtained from Bisphenol A. Resins with **n** values from 0 to 30 are commercially available, although **n** values greater than 12 are not normally employed in surface coating systems.

Epoxy resins are characterised by their Epoxide Equivalent Weight (E.E.W.) which is defined as, The Weight of Resin containing one gram equivalent of Epoxide. The E.E.W. is also referred to as, Weight Per Epoxide (W.P.E.) or Epoxy Molar Mass (E.M.M.). All three terms are interchangeable.

n Value	Molecular Weight	E.E.W.	Melting Point °C
0 to 1	350 − 600	170 − 310	<40°C*
1 to 2	600 − 900	310 − 475	40 − 70°C
2 to 4	900 − 1400	475 − 900	70 − 100°C
4 to 9	1400 − 2900	900 − 1750	100 − 130°C
9 to 12	2900 − 3750	1750 − 3200	130 − 150°C

*Usually liquid at ambient temperature

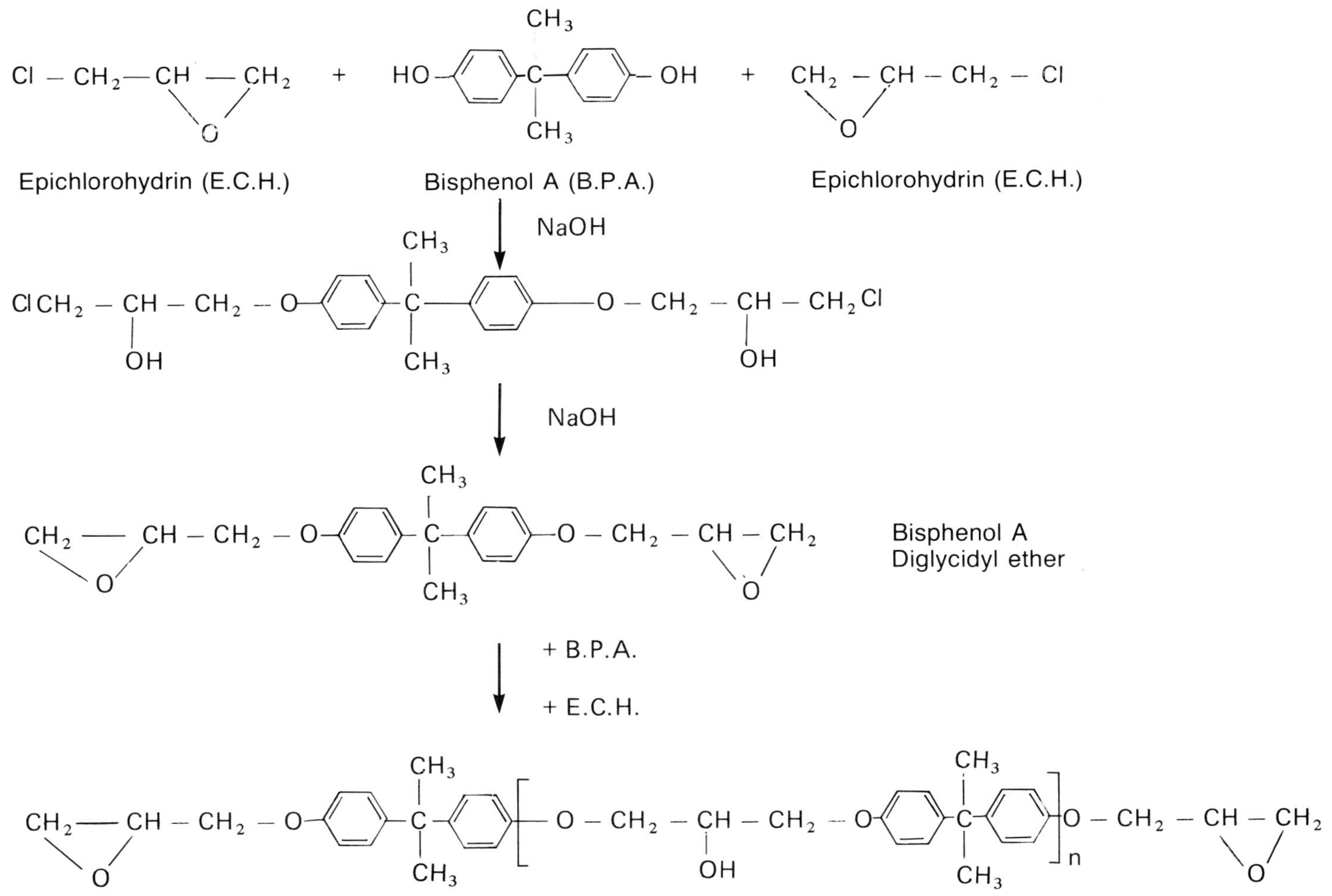

5

CI — CH$_2$ — CH — CH$_2$ (C, O triangle) + HO — (ring) — C(CH$_3$)(CH$_3$) — (ring) — OH + CH$_2$ — CH — CH$_2$ — CI (O triangle)

Epichlorohydrin (E.C.H.)   Bisphenol A (B.P.A.)   Epichlorohydrin (E.C.H.)

NaOH

CICH$_2$ — CH — CH$_2$ — O — (ring) — C(CH$_3$)(CH$_3$) — (ring) — O — CH$_2$ — CH — CH$_2$ CI
                    |OH |OH

NaOH

CH$_2$ — CH — CH$_2$ — O — (ring) — C(CH$_3$)(CH$_3$) — (ring) — O — CH$_2$ — CH — CH$_2$
(O triangle)   (O triangle)

Bisphenol A
Diglycidyl ether

+ B.P.A.
+ E.C.H.

CH$_2$ — CH — CH$_2$ — O — (ring) — C(CH$_3$)(CH$_3$) — (ring) — [O — CH$_2$ — CH — CH$_2$ — O — (ring) — C(CH$_3$)(CH$_3$) — (ring)]$_n$ O — CH$_2$ — CH — CH$_2$
(O triangle) |OH   (O triangle)

Although the molecular structure given above suggests that all epoxy resins are difunctional, with respect to the epoxide group, this is not entirely true for commercial epoxy resins. Other end groups are present as a consequence of the manufacturing process, e.g. unreacted phenolic OH groups, unconverted chlorohydrin groups, and glycols formed by the breakdown of the epoxide group. Low molecular weight resins have a functionality of >1.9, but, as the molecular weight increases, the functionality of the epoxy resin approaches one, as illustrated in the table on page 4.

CURING OF EPOXY RESINS

The most important curing reactions used to form coatings are detailed below.

Polymerisation via the Epoxide Group

The epoxide group is readily ring opened by tertiary amines to give large molecules linked via ether groups.

$$R_3N^+ - CH_2 - CH \sim + \quad CH_2 - CH \sim \longrightarrow R_3N^+ - CH_2 - CH \sim$$

with the first CH bearing O^-, the epoxide CH_2-CH sharing O, and the product chain showing $O-CH_2-CH\sim$ with terminal O^-.

When OH groups are present in the epoxy resin, then a further reaction is possible. The tertiary amine can induce alkoxide ion formation and hence provide a site for chain branching.

$$\sim\!CH - CH_2 \xrightarrow{\ R_3N\ } \sim\!CH - CH_2$$

with OH and epoxide O on the left, converting to O^- and epoxide O on the right.

reacting with $CH_2-CH \sim$ (epoxide)

giving:

$$\sim CH - CH_2$$
$$\quad | \qquad \backslash\!/$$
$$\quad O \qquad O$$
$$\quad |$$
$$\quad CH_2$$
$$\quad |$$
$$\quad CH - O^-$$

Thus a complex, three dimensional, high molecular weight structure is achieved when both reactions occur during epoxy resin cure.

$$CH_2 - CH \sim\sim\sim\sim\sim\sim\sim\sim\sim\sim\sim\sim\sim CH - CH_2$$

(structure)

```
CH₂– CH ~~~~~~~~~~~~~~~~~~~~~~~~~~~~ CH – CH₂
   \ /                    |              \ /
    O                     O               O
                          |
                          CH₂
                          |                    O⁻
                          CH – O -- CH₂——CH
              HO —————|                 |———OH
                   O⁻—— CH              CH
                          |            / \
                          |           O
                          CH₂          \
                          |             CH₂
```

Reactions with Primary and Secondary Amines

Primary aliphatic amines react at room temperature, with the epoxide ring forming a secondary amine, which in turn reacts with a further epoxide ring, forming a tertiary amine link between two epoxy resin molecules.

```
                              R
                              |
~~~~CH – CH₂      +      H – N – H
      \ /
       O
       |
       ↓

                              R
                              |
        ~~~~CH – CH₂ – N – H      +      CH₂ – CH~~~~
             |                                \ /
             OH                                O
                                               |
                                               ↓

                              R
                              |
            ~~~~CH – CH₂ – N – CH₂ – CH~~~~
                 |                    |
                 OH                   OH
```

The reaction of the hydroxyl groups, so formed, with epoxide groups is suppressed in the presence of secondary and primary amines. The tertiary amines formed are too sterically hindered to act as a catalyst for the hydroxyl-epoxide reaction shown above.

Since each primary amine is difunctional with respect to epoxy resins, the use of primary diamines (such as ethylene diamine) will result in a highly cross linked structure being formed.

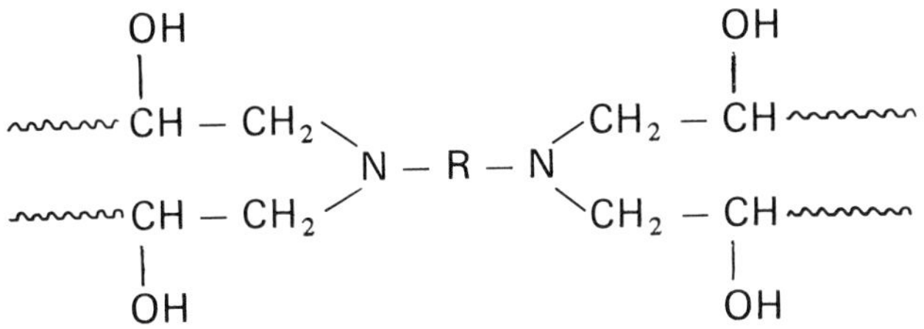

In practice, primary aliphatic diamines are too volatile to be satisfactorily used as curing agents, but higher molecular weight polyamines (usually diamine dimer acid adducts) are used as 'two pack' systems. Aromatic amines are not as reactive as aliphatic amines. They usually require temperatures of about 200°C before significant reaction can occur. As a result, 'one pack' systems with long shelf lives at room temperature can be obtained from aromatic amine/epoxy resin combinations.

Reaction with Acid Anhydrides

The reaction of carboxylic acids with epoxy resins is discussed under the section dealing with epoxy esters.

The use of polybasic carboxylic acids as a method of curing non ester coatings is of little commercial importance. However, the use of acid anhydrides is second in importance only to amine curing agents.

The following reaction sequence occurs:

a) The anhydride is ring-opened to give the half-ester.

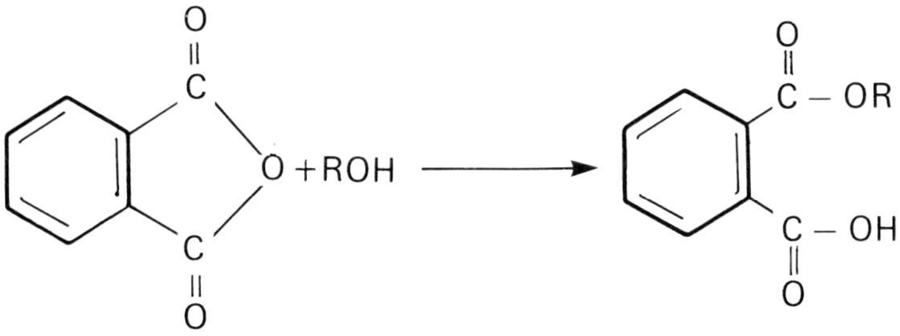

With commercial epoxy resins, this step can be brought about initially by water or hydroxyl compounds present in the resin.

b) The carboxyl group formed in (a) reacts with an epoxide group to form a hydroxy diester:

c) The hydroxy group formed in (b) can then either ring open another anhydride group:-

or react with an epoxide group to form an ether link. The presence of the acid anhydride, acts as a catalyst for this reaction.

The curing reaction between acid anhydrides and epoxy esters is slow if uncatalysed, even at a temperature of 200°C. In the presence of basic or acid catalysts, the reaction proceeds readily. Basic catalysts, and tertiary amines in particular, favour the formation of diester structures.

Reaction with Methylol and Methylol Ether Groups

Methylol and methylol ether groups react with epoxy resins via (a) the epoxide group and (b) any hydroxyl groups present on the epoxy resin molecule (see a) and b) below).

In addition any methylol and methylol ether groups are capable of self reaction. Thus when amino or phenolic resins are used to co-cure with epoxy resins, the result is a highly complex cross linked film.

Since the reaction with amino or phenolic resins is very slow at room temperature, stable 'one-pack' systems can be obtained. Reaction between the epoxy resin and the amino or phenolic resin is initiated by the use of acidic catalysts or by elevated temperatures (over 100°C).

a)

b)

Where R^1 is H or an alkyl group

Curing Conditions

Epoxy resin coatings obtain their excellent properties through reaction with curing agents.

The curing agents react with the epoxide groups and/or hydroxyl groups of the epoxy resin, to give stable carbon-carbon, carbon-oxygen, or carbon-nitrogen links.

It is these stable linkages that confer the epoxy resin film with, among other properties, its excellent chemical and solvent resistance.

With a correct choice of cross linking agent, resin and modifier, the properties of epoxy resin films can be tailored to cover a wide variety of performance characteristics.

Curing agents for epoxies can be divided into two categories, i.e. those that give cured films at ambient temperature, and those that require elevated temperature (baking or stoving) to cure.

Ambient temperature cure is predominantly carried out using aliphatic polyamines, polyamine adducts or polyamides. Because of their reactivity at ambient temperature, they are sold as 'two-package' systems, and the amine is kept apart from the epoxy resin until just before use. Pot-life for these systems varies from a few minutes to several days, and is dependent upon the epoxy resin type, amine type, solvent type and also on storage conditions. Most commercial systems are compounded to give a pot-life between 8 to 12 hours.

Polyamine types in general use include ethylene diamine, diethylene triamine, triethylene tetramine and tetraethylene pentamine. Several cycloaliphatic amines are also used commercially. They all offer low viscosity, rapid cure at low temperature and excellent chemically resistant film properties.

It should however, be noted that unmodified amines present handling problems, because of their corrosive nature.

Less hazardous systems are obtained by using reactive polyamines and polyamine adducts in place of the amines. Cure rates are generally slower than with unmodified amines − but the resulting films are tougher and more flexible. The use of reactive polyamides probably gives the best all round properties. Film strength, impact resistance, adhesion, gloss retention and flexibility are all generally superior. However, chemical resistance and solvent resistance tend to be inferior to other low temperature cure systems.

Heat cure involves the hydroxyl and epoxide groups of the epoxy resin. Aromatic amines, acid anhydrides, amino resins and phenolic resins are used in commercially available heat cure systems. Although they require elevated temperature for cure, they have the advantage of providing 'one-package' systems that are stable at ambient temperature.

Aromatic amines in general use include 4, 4′ methylene dianiline and m-phenylene diamine. They require temperatures of up to 200°C for complete cure and impart excellent physical and chemical resistance properties to the cured film.

Anhydrides are the most commonly used curing agents, after polyamines. Those used commercially include phthalic anhydride, trimellitic anhydride, hexa-hydro phthalic anhydride and methyl hexahydrophthalic anhydride (nadic anhydride). In general they require higher curing temperatures than other types of heat cure catalysts, but have longer pot lives. Chemical resistance is inferior to polyamine and amino resin cured systems, but generally heat distortion temperatures of the films are superior.

Amino and phenolic resins are co-cured with epoxy resins to obtain extremely hard, flexible, chemically resistant films. In addition to reacting with the epoxy resin, these materials cross-link with themselves and the final film properties are a combination of the epoxy resin and amino resin film characteristics. Types representing the whole spectrum of melamine-formaldehyde, urea-formaldehyde and phenol-formaldehyde resins are used commercially to co-cure with epoxy resins.

All types require the addition of catalytic amounts of acid to initiate cure (e.g. paratoluene sulphonic acid or benzoic acid). The temperature required for cure varies with resin type but is normally greater than 100°C.

In general amino resins, and in particular melamine types, require lower cure temperatures than phenolic resins. They also result in paler coloured films but are more expensive than phenolic resins. Urea types are more commonly used in non-ester coatings, while melamine types are more common in epoxy ester systems. Phenolic resins tend to provide cured films with greater heat stability and superior chemical and solvent resistance.

The amount of curing agent required, is determined by the type and stoichiometry of the chemical reaction. Where a cross linking reaction is required, there is a calculable level of curing agent, which is necessary for complete reaction of all available functional groups in the epoxy resin. In practice, the actual level is optimised for the desired performance properties of the film, and very small variations in the quantity of curing agent used can have major effects on film properties.

When the curing agent is catalytic in its action (e.g. tertiary amines), the exact amount required must be arrived at empirically for each system and may be varied widely without significant effects on film properties. When calculating amounts of curing agent the following may be taken as a guide:

1 amino hydrogen per epoxide equivalent weight of epoxy resin

1 Carboxyl group per epoxide equivalent weight of epoxy resin

½ Anhydride group per epoxide equivalent weight of epoxy resin.

All the normally available range of pigments can be used. However, alkaline pigments and metal pigments should be avoided when cure is by acid or acid containing systems.

As might be expected, the inclusion of pigments and fillers lengthens the cure times required.

The use of solvents allow systems to be prepared at viscosities which can easily be handled. The solvent acts as a carrier for the resin and allows a continuous smooth film to be applied. After appliction, it is essential that the solvent should completely evaporate from the coating. If solvent release is incomplete the film properties are adversely affected.

Epoxy resins are usually soluble only in highly polar solvents such as ketones, esters and ethers. These are amongst the more expensive solvents available to the coatings industry. The inexpensive solvents such as xylene, toluene, white spirit, etc are poor solvents for epoxy resins. To reduce cost, most solvent systems are mixtures of expensive true solvents (e.g. methyl isobutyl ketone, Cellosolve acetate, etc.) and inexpensive diluents (e.g. xylenes).

When selecting a solvent system the other components of the system must be taken into consideration. For example, the use of esters, such as cellosolve acetate, in an amine cure system results in solvent-amine interaction which can considerably inhibit the curing reaction.

Environmental pressures are enforcing changes in the solvents employed with epoxy resin systems and resins have been developed for use as 100% solid systems and as water borne systems.

APPLICATIONS FOR TWO-PACK EPOXY COATINGS

Epoxy resins have found commercial applications in the field of high performance protective and decorative finishes, despite the fact that they are more expensive than many other surface coating materials. This is because of the excellence of their performance properties particularly in areas where high chemical resistance and adhesion are required.

Low temperature cure two package systems, offer resistance properties normally obtainable only with stoving systems. They are employed extensively in heavy duty industrial maintenance coatings, industrial floor coatings, marine maintenance coatings and tank linings.

Phenolic cure systems are used in beverage and food can coatings; as drum and pipe linings; in wire coatings, and as impregnating varnishes.

While not as chemically resistant as phenolic cure systems, amino resin systems are used where good colour is required and are extensively used in coatings for can linings, appliance primers and metal furniture topcoats.

Formulations

The following formulations are typical.

AMINE CURE SYSTEM

Epoxy resin (E.E.W. 450)	52.0
Flow control agent	2.9
n-Butanol	9.8
Butyl Cellosolve	1.3
Methyl ethyl ketone	10.2
Xylol	21.2
Diethylene triamine	2.6
	100.0

POLYAMIDE CURE SYSTEM

Epoxy resin (E.E.W. 500)	20
Polyamide (amine value 90 mgKOH/g) (60% non-volatile content)	50
Methyl ethyl ketone	15
Xylol	15
	100.0

UREA FORMALDEHYDE CURE SYSTEM

Epoxy resin (E.E.W. 1500)	21
Urea formaldehyde resin (70% non-volatile content)	30
Xylol	17
n-Butanol	17
Methyl isobutyl carbinol	15
	100.0

Cure schedule: 30 minutes at 180°C

POWDER COATINGS

Powder coatings have been the major growth area for epoxy resins during the last 10 years.

Pigments, flow additives and a curing agent are dispersed quickly into molten epoxy resin. This melt is cooled to a solid, as quickly as possible, in order to quench the reaction of the curing agent. The pigmented solid is fairly stable at ambient temperatures and can be ground to a powder and applied to a substrate without cross linking.

The powder can be applied by electrostatic spray to a cold or a hot metal object.

Alternatively, the object may be pre-heated and then dipped into the fluidised powder. The object is then baked at 170 to 200°C so that the pigmented powder will melt and flow to form a smooth coating. At this temperature, the curing agent will cross link the epoxy resin. Dicyandiamide is a suitable curing agent.

Epoxy resins may be modified with polyester resins, to give powder coatings with improved gloss retention during exterior exposure. This type is often cured with triglycidy-lisocyanurate.

Epoxy polyester powders can also be suspended in water and applied by conventional techniques.

Powder and water dispersible coatings are considered in detail in Volume III.

WATER DISPERSIBLE EPOXY COATINGS

Water based systems have been developed, to satisfy the demand for the replacement of organic solvents by water for safety and pollution control. In addition to aqueous powder suspensions, water based versions of two pack and single pack systems exist, and, of course, epoxy resins are very important constitutents of aqueous electrodeposition coatings.

Liquid epoxy resins may be emulsified, using external emulsifiers, or by internal modification of the resins. Some curing agents, such as polyamides, can be used as the emulsifier. Aqueous two pack systems are often chosen for maintenance coatings in areas where solvent vapour would be a problem, e.g. hospitals, abattoirs and badly ventilated areas, such as inside the hull of a ship.

A major use for single pack aqueous coatings is as an interior coating for cans.

High molecular weight epoxy suspensions are cured with aqueous melamine resin or phenolic resin curing agents.

EPOXY ESTER RESINS

The majority of epoxy resin esters are the reaction products of an epoxy resin and a vegetable oil fatty acid. They form a range of products and like alkyds they are characterised by oil length and oil type. Both air drying and stoving types are used commercially.

Although more expensive and less versatile than alkyds, epoxy esters generally have better colour, flexibility, adhesion and chemical resistance.

Preparation

The main chemical reactions occuring during epoxy ester preparation are:

ESTERIFICATION

i) $\quad -\overset{\displaystyle ||}{\underset{\displaystyle O}{C}}-OH \quad + \quad \underset{\displaystyle O}{CH_2-CH}\sim\!\sim\!\sim \quad\longrightarrow\quad -\overset{\displaystyle ||}{\underset{\displaystyle O}{C}}-OCH_2-\overset{\displaystyle }{\underset{\displaystyle OH}{CH}}\sim\!\sim\!\sim$

Hydroxy ester

ii) $\quad -\overset{\displaystyle ||}{\underset{\displaystyle O}{C}}-OH \quad + \quad -\overset{\displaystyle }{\underset{\displaystyle OH}{CH}}- \quad\longrightarrow\quad -\overset{\displaystyle }{\underset{\displaystyle O-\overset{||}{\underset{O}{C}}-}{CH}}- \quad +H_2O$

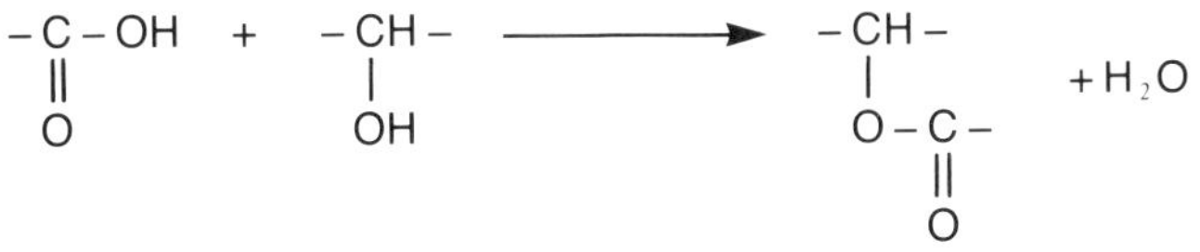

Condensation ester

ETHERIFICATION

iii) $\quad -\overset{\displaystyle }{\underset{\displaystyle OH}{CH}}- \quad + \quad \underset{\displaystyle O}{CH_2-CH}\sim\!\sim\!\sim \quad\longrightarrow\quad -\overset{\displaystyle }{\underset{\displaystyle OCH_2-\overset{}{\underset{OH}{CH}}\sim\!\sim\!\sim}{CH}}-$

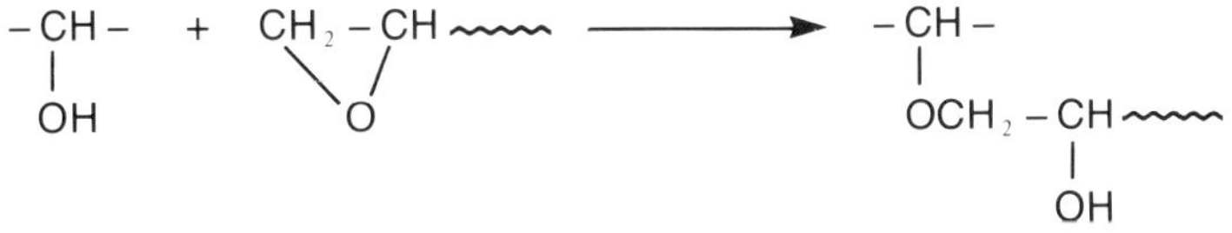

Hydroxy ether

During the reaction between the fatty acids and epoxy resin, the acid number of the system falls as the viscosity increases.

Reactions I and II result in only a small viscosity increase for a correspondingly large decrease in acid number.

In order to produce commercially desirable low viscosity, low acid number esters, it is necessary to suppress reaction III whilst encouraging reactions I and II.

Efficient water removal promotes reaction II, and this can be aided by the use of xylene, or other suitable reflux solvents, to azeotropically remove water of reaction. The most effective method of controlling the relative amounts of esterification to etherification, however, is by use of an esterification catalyst.

Suitable catalysts include alkalis, basic organic compounds and organometallic salts, e.g.

Sodium or lithium carbonate

Tertiary amines

Triphenyl phosphine

Zirconium or stannous octoate.

Sodium and lithium are the most specific for the COOH-epoxide group reaction, while zinc and zirconium promote esterification but tend to favour the COOH − OH reaction in preference to the COOH − epoxide reaction.

Catalyst levels employed are usually of the order of 0.01−0.05%, based on the solid epoxy resin.

The viscosity of an epoxy ester is dependent upon the ester specificity of the catalyst employed.

In the context of epoxy esters, the epoxy resin can be considered as a polyfunctional high molecular weight alcohol. Although esters can be prepared from all the commercially available types, epoxy resins with E.E.W.'s in the range 700−1000 are normally used.

Some grades of epoxy contain amounts of esterification catalyst as part of the manufacturing process.

All the vegetable oil fatty acids common to alkyd resin manufacture are also used in epoxy-ester manufacture. The most commonly used are listed below together with the properties they confer on the epoxy ester.

Fatty Acid	Epoxy Ester Properties
Linseed	Fast air drying systems with poor colour retention
Dehydrated Castor	Fast air drying or stoving systems with good flexibility and chemical resistance
Soya Bean	Air drying systems with good colour and soft flexible films
Coconut	Non air drying systems with very good colour, chemical resistance and flexibility. Good colour on overbaking.

Complete esterification involves the use of one equivalent of fatty acid to one epoxy resin esterification equivalent. However, in order to obtain a low acid number product at a handleable viscosity, most commercial esters are formulated with less fatty acid than the calculated theoretical amount. Where the epoxy ester is intended for co-cure (e.g. with amino resins) the excess epoxy resin must be large enough to ensure sufficient residual reaction sites are available for curing.

In practice the amount of fatty acid used in epoxy ester manufacture varies from 30% to 90% of the esterification equivalent of the epoxy resin as illustrated below:

Epoxy Resin Esterification Equivalent	Fatty Acid Equivalent	Fatty Acid Content (wt. % calculated on E.E.W. of 900)	Oil Length
1.0	0.3−0.5	30−40	Short
1.0	0.5−0.7	40−50	Medium
1.0	0.7−0.9	50−55	Long

The effect of the epoxy resin fatty acid ratio on the film properties of the epoxy ester can be summarised, as follows:

Chemical Resistance Film Hardness Drying Time Adhesion Gloss retention	All increase with decreasing fatty acid content.

Flow Solubility in aliphatic hydrocarbon solvents Water Resistance Flexibility Pigment wetting Exterior durability Colour	All increase with increasing fatty acid content

In addition, the viscosity and cost of the epoxy ester, decrease with increasing fatty acid content.

Dimer acid, rosin acid, benzoic acid or unsaturated monomers, such as vinyl toluene and styrene, are often used as modifiers.

Dimerised fatty acids are frequently included to improve the flexibility and water resistance of epoxy esters. Levels of between 2% and 6% of the fatty acid content are typical, however, care must be exercised in deciding the level of dimer acid used in any particular formulation. Its inclusion leads to chain extension, and excessive amounts result in high molecular weight, high viscosity esters.

Rosin acids impart faster drying characteristics and increased film hardness at the expense of colour, flexibility and chemical resistance.

Incorporation of small quantities of benzoic acid give improved film hardness and mar resistance but can reduce flexibility. Styrene or vinyl toluene modification is frequently used to improve colour, film hardness and drying properties. Vinyl monomers are introduced by one of two methods.

i) **Post vinylation:** The epoxy ester of a conjugated fatty acid (e.g. D.C.O.) is first prepared, then vinylated in a separate operation, utilising the unsaturation of the fatty acid as sites for copolymerisation.

ii) **Prepolymer method:** A copolymer of the vinyl monomer and 2−4% of an acid monomer (e.g. methacrylic acid) is prepared, then reacted on to an epoxy resin by reaction of the acid groups with some of the epoxide groups. Finally, fatty acid is added and reacted with the epoxy resin to form the epoxy ester.

The degree of copolymerisation obtained between ester and vinyl monomer, using method (i), is relatively low, and a large amount of the vinyl species is present, as a homopolymer. On storage, this homopolymer can precipitate, resulting in the formation of hazy films on application. Both post vinylation and prepolymer methods result in resins with similar performance characteristics.

EPOXY ESTER MANUFACTURE

Epoxy esters are manufactured using similar equipment to that used for alkyd resin manufacture, i.e. a stainless steel reactor fitted with condenser, Agitator, separator (water trap) and inert gas supply. Esterification is normally carried out under solvent reflux.

Although epoxy esters can be prepared by fusion techniques, the reaction time is increased and the final product exhibits higher viscosity and darker colour than when solvent processing is used. The presence of alkaline or acidic contaminants (e.g. raw material residues from a previous preparation) can cause unwanted side-reactions to occur, which effect the final epoxy ester viscosity.

The course of the reaction is followed by sampling from the reactor at hourly intervals and determining acid number and viscosity. Most epoxy esters are reacted to acid values of below 10mgKOH/g and often below 1mgKOH/g. Process time varies with the catalyst used but is normally 4−10 hours.

TYPICAL EPOXY ESTER FORMULATIONS

RESIN A

**AIR DRYING EPOXY ESTER
— MEDIUM OIL LENGTH**

Epoxy resin (E.E.W. 950)	27.9
DCO Fatty acid	22.8
Zirconium octoate	(0.06% Zr metal on epoxy resin charge)
Xylol (reflux solvent)	2.6
Xylol (thinning solvent)	46.7
	100.0

PROCESS — Total process time approximately six hours.

1. Charge DCO fatty acid and zirconium octoate to the reactor. Start agitator and blanket reactor with inert gas sweep.
2. Heat to 150°C.
3. Add the epoxy resin at a rate that does not foul the agitator while maintaining the temperature at 150°C.
4. When all the epoxy resin has been added and a homogeneous solution has been formed set reactor for solvent process.
5. Add Xylol reflux solvent and heat to 240°C.
6. Maintain the temperature at 240°C collecting water of reaction in the separator (water trap) and returning reflux solvent to the reactor.
7. Sample at hourly intervals and determine the acid number.
8. When acid number falls to below 3mgKOH/g cool to below 100°C and add Xylol thinning solvent.

TYPICAL RESIN CONSTANTS

Colour	5 Gardner
Non-volatile content	50%
Acid number	<3 mgKOH/g
Viscosity (Gardner-Holdt) (at 25°C)	V – W

RESIN B

EPOXY ESTER — MELAMINE STOVING SYSTEM — MEDIUM OIL LENGTH

Epoxy resin (E.E.W. 800)	30.6
Coconut fatty acid	25.1
Sodium benzoate	(0.03% Na on epoxy resin charge)
Xylol (reflux solvent)	2.3
Xylol (thinning solvent)	42.0
	100.0

PROCESS — As for resin A but conduct esterification at 250°C.
Total process time approximately six hours.

TYPICAL RESIN CONSTANTS

Colour	5 Gardner
Non-volatile content	55%
Acid Number	<3 mgKOH/g
Viscosity (Gardner-Holdt) (at 25°C)	Y – Z

RESIN C

EPOXY ESTER — STOVING SYSTEM — MEDIUM OIL LENGTH

Epoxy resin (E.E.W. 850)	30.6
Tall oil fatty acid	22.6
Dimer acid	2.5
Sodium benzoate	(0.02% Na on epoxy resin charge)
Xylol (reflux solvent)	2.3
Xylol (thinning solvent)	42.0
	100.0

PROCESS — As for resin B.
Total process time approximately eight hours.

TYPICAL RESIN CONSTANTS

Colour	6 Gardner
Non-volatile content	55%
Acid Number	<1 mgKOH/g
Viscosity (Gardner-Holdt) (at 25°C)	V – W

RESIN D

EPOXY ESTER — LONG OIL LENGTH

Epoxy resin (E.E.W. 650)	24.4
Tall oil fatty acid	17.9
Gum rosin	7.7
Zirconium octoate	(0.05% Zr on epoxy resin charge)
Xylol (reflux solvent)	2.6
Xylol (thinning solvent)	47.4
	100.0

PROCESS — As for resin B.

Total process time approximately 10 hours.

TYPICAL RESIN CONSTANTS

Colour	6 – 7 Gardner
Non-volatile content	50%
Acid Number	7 – 10 mgKOH/g
Viscosity (Gardner-Holdt) (at 25°C)	V – W

APPLICATIONS OF EPOXY ESTER COATINGS

Although epoxy esters have similarities with alkyd resins, they offer films with some superior film properties. In particular:

Adhesion

Flexibility

Chemical resistance

Colour.

The overall properties are inferior to two-pack epoxy resin films but they can have some advantages:

Lower cost

Fast stoving schedules

Better pigmentation properties

Long shelf life.

They combine some of the ease of handling and application of alkyd resins, with some of the film properties of epoxy resins. They are used for air-dry or stoving coatings or they can be co-cured with phenolic or amino resins. Application can be by roller coating, spraying, brushing or electro-deposition. Epoxy esters can be used as:

Automotive primers

Appliance primers

Flexible tube coatings

Drum linings

Marine finishes

Floor sealers and top coats

Metal decorating lacquers

Enamels for hardware and metal furniture

Industrial maintenance primers and topcoats.

TYPICAL COATING FORMULATIONS

AIR DRY ZINC RICH PRIMER FOR METALS

Epoxy ester resin A (50% non-volatile content in Xylol)	12.9
Zinc dust	79.7
Xylol	4.8
Shellsol A	1.6
Bentone	0.8
Cobalt naphthenate (6% Co)	0.05
Calcium naphthenate (4% Ca)	0.15
	100.0

Non-volatile content	87%
Pigment volume concentration	65%

EPOXY ESTER — MELAMINE STOVING
SYSTEM FOR APPLIANCE PRIMER

Epoxy ester resin B (55% non-volatile content in Xylol)	59.0
Titanium dioxide	17.0
Lampblack	0.5
Melamine resin (66% non-volatile content)	23.5
	100.0

Non-volatile content	64%
Pigment volume concentration	9%

The pigment is dispersed in the epoxy ester prior to adding the melamine resin.

The formulation is reduced to spraying viscosity with Xylol and cured for 30 minutes at 175°C after application.

EPOXY ESTER FOR USE
AS AUTOMOTIVE PRIMER

Epoxy ester resin C (55% non-volatile content)	30.8
Red iron oxide	9.8
Yellow iron oxide	2.4
Black iron oxide	4.9
Barium sulphate	14.8
Zinc chromate	0.9
Bentone 34	0.8
Xylol	35.5
Manganese naphthenate (6% Mn)	0.1
	100.0

Non-volatile content	50%
Pigment volume concentration	32%

Pigment dispersion is carried out with only 25% of the quantity of xylol shown above. The formulation may be spray applied and stoved for 40 minutes at 140°C to effect cure.

AUTOMOTIVE PRIMER STOVING SYSTEM

Epoxy ester resin D	25.0
Urea formaldehyde resin	2.5
Red iron oxide	8.0
Barium sulphate	11.1
Talc	6.3
China clay	6.3
Aluminium	0.6
Xylol	40.2
	100.0

Cure schedule: 40 minutes at 135°C

Chapter VI

Phenol-Formaldehyde and Amino Resins

28

Chapter VI

Phenol-Formaldehyde and Amino Resins

Amines, amides and phenols can be reacted with formaldehyde to give resinous products suitable for use in surface coating systems. Although there are points of similarity (particularly in their end uses) between amino resins and phenolic resins, the two types will be considered separately in this chapter.

PHENOL-FORMALDEHYDE RESINS

Phenolic resins are the polymeric products of a controlled reaction between a phenol and formaldehyde. They have a wide variety of commercial applications including those for laminates, wood adhesives, moulding powders and cavity insulation. In the surface coating field they can be used on their own or as modifiers with other resins. In this latter application they are often used to provide sites for the cross linking or curing of otherwise thermoplastic polymer species. Phenolic resins are produced by a two stage reaction.

Methylolation

The reaction between phenol and formaldehyde results in electrophillic substitution of the aromatic ring to form a phenol alcohol. This is termed methylolation.

The sites at which methylolation occurs are dependent upon the electron density at the substitution points on the ring and are dictated by both the nature and position of any substituent group already present.

The phenolic OH group enhances the electron density of the two, four and six positions of the aromatic ring. Substitution is thus directed towards the ortho and para positions. When the ortho and para positions are blocked by electron releasing groups, some substitution may occur, to a very limited extent, in the meta positions.

The presence of electron withdrawing groups on the ring (e.g. NO_2) results in the suppression of the methylolation reaction.

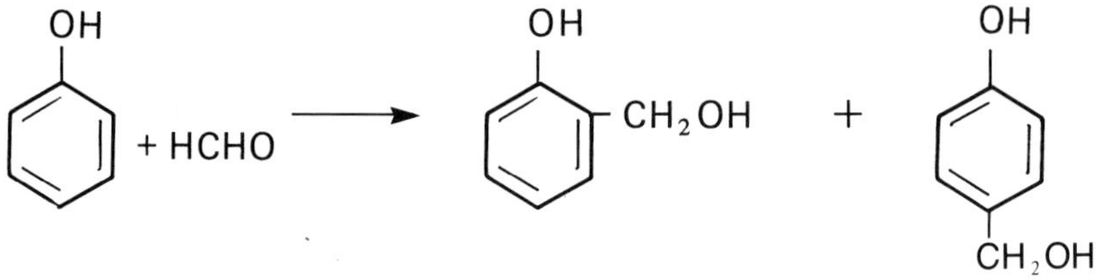

The above depicts an idealised reaction. In practice di- and tri-methylol phenols will also be present. The exact nature of the product mix will depend on the relative proportions of formaldehyde and phenol present and on the reaction conditions.

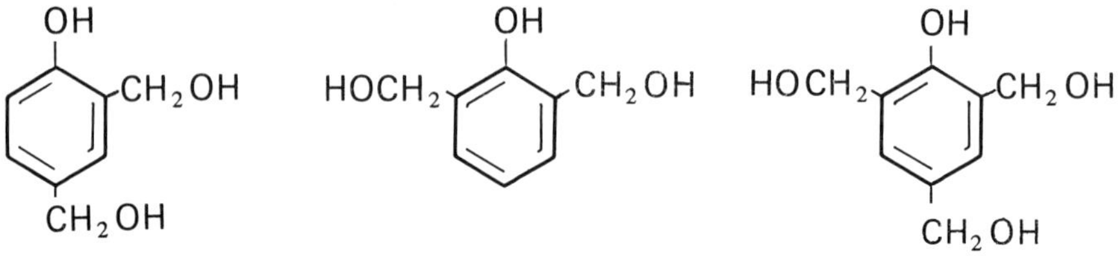

2, 4 di-methylol phenol   2, 6 di-methylol phenol   2, 4, 6 tri-methylol phenol

Condensation

Methylol phenols are relatively stable at temperatures below 50°C and high (alkaline) pH. However at higher temperatures, or under the influence of acid catalysts, reaction takes place between the methylol groups of neighbouring molecules to give polymeric products.

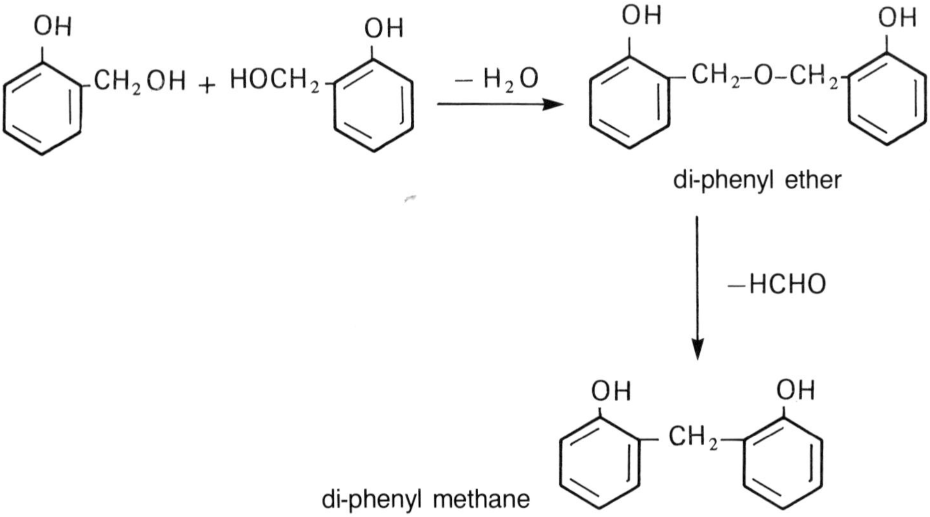

di-phenyl ether

$-HCHO$

di-phenyl methane

Depending on the reaction conditions, a number of polynuclear structures can be formed. For example where di- and tri-methylol phenols are present branched polymer chains can be formed.

The mole ratio of formaldehyde: phenol and the conditions under which they are reacted, have a pronounced effect on the relative rates at which the methylolation and polymerisation reactions proceed and thus on the structure of the product formed.

Alkaline conditions coupled with F:P ratios (mole ratio formaldehyde:phenol) of greater than 1:1 favour the methylolation reaction, while acidic conditions together with F:P ratios of less than 1:1 favour condensation. The pronounced effect of the F:P ratio on the resin produced, gives rise to the two basic types of phenolic resin encountered in surface coating systems, the resole and the novolac.

Phenol + Formaldehyde (excess) — Alkaline catalyst → Resole

Phenol + Formaldehyde (excess) — Acid catalyst → Novolac

PHENOL-FORMALDEHYDE RESOLES

A resole is a phenolic resin which, when heat is applied, will cross link to form a solid, insoluble, infusable, three-dimensionally linked polymer network.

Resoles are prepared using an F:P ratio of greater then 1:1. Reaction between formaldehyde and phenol is brought about under highly alkaline conditions, pH 8.5 or above being commonly used. These reaction conditions maximise the amount of methylolation giving rise to the formation of methylol phenols with virtually no condensation occurring.

The introduction of the first methylol group onto the benzene ring increases the likelihood of further methylolation of the ring.

Because of this fact some trimethylol phenols will be formed, even when the overall F:P ratio is considerably less than 3:1.

For this reason when dimethylol phenols are required it is usually more effective to use a phenol derivative with a blocked para position (e.g. para tertiary butyl phenol) rather than to try to form the dimethylol phenol by control of the starting F:P ratio, as illustrated by the following examples:

PHENOL

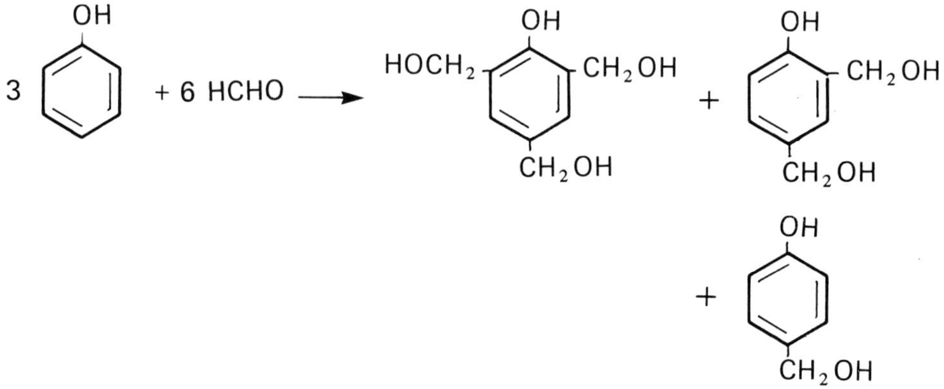

p-TERTIARY BUTYL PHENOL

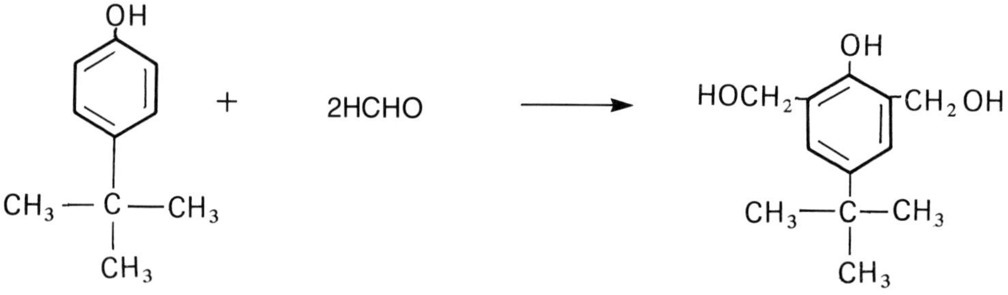

Polymerisation of the methylol phenol to form a low molecular weight polymer may be brought about by a condensation reaction under either alkaline or acidic conditions.

The course of the reaction varies with the reaction conditions as illustrated below.

THERMAL CONDITIONS

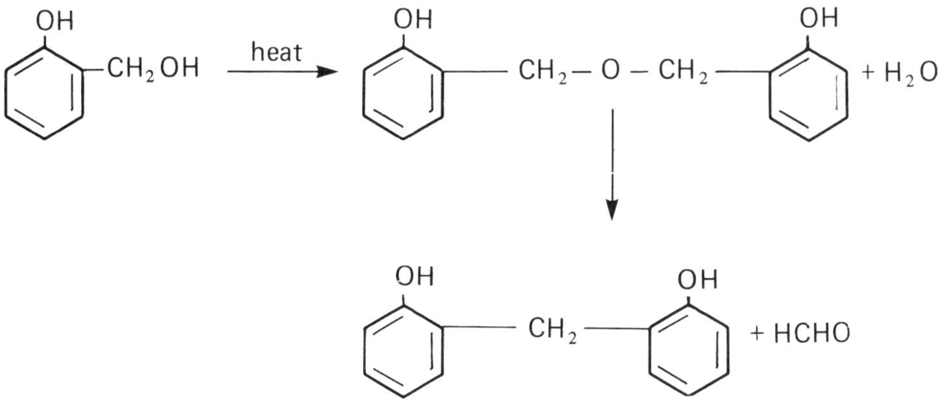

OH—CH₂OH →(heat) OH—CH₂—O—CH₂—OH + H₂O

↓

OH—CH₂—OH + HCHO

ACIDIC CONDITONS

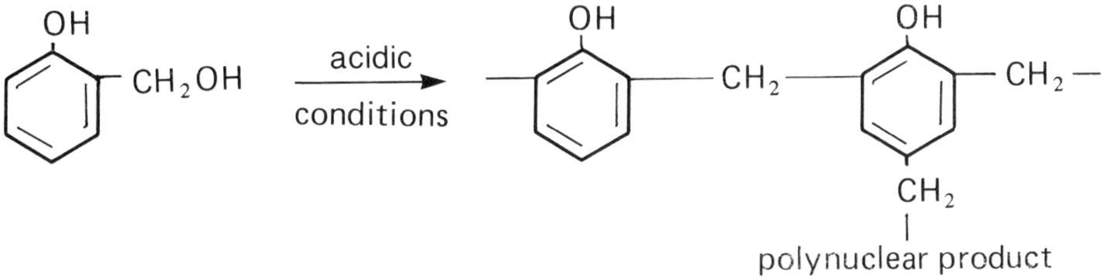

OH—CH₂OH →(acidic conditions) —OH—CH₂—OH—CH₂—
with CH₂ branch
polynuclear product

ALKALINE CONDITIONS

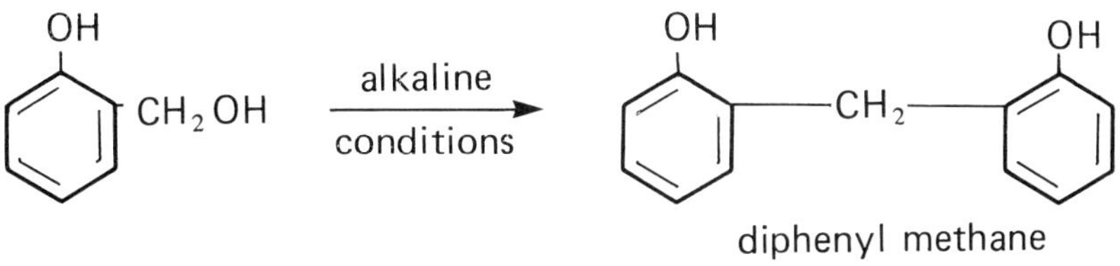

OH—CH₂OH →(alkaline conditions) OH—CH₂—OH

diphenyl methane

Thus commercial resoles are a mixture of polynuclear molecules joined via methylene or methoxy links and containing residual unreacted methylol groups.

THE CURING REACTIONS OF PHENOLIC RESOLES

Heat curing of a resole may occur via the quinone methide formation.

Quinone methide

Acid cured resoles are used commercially as binders, the main acid catalysts used are p-toluene sulphonic acid, hydrochloric acid or phosphoric acid.

In addition to the self cure reactions which are outlined above the residual methylol groups on the resole, allow reaction with other polymer species containing hydroxyl, carboxyl, amino or epoxy groups. These reactions are outlined in detail under the section dealing with amino resin co-cure and are essentially the same for amino and phenolic resins.

A reaction of particular importance in surface coating applications is the reaction between phenolic resoles and unsaturated compounds.

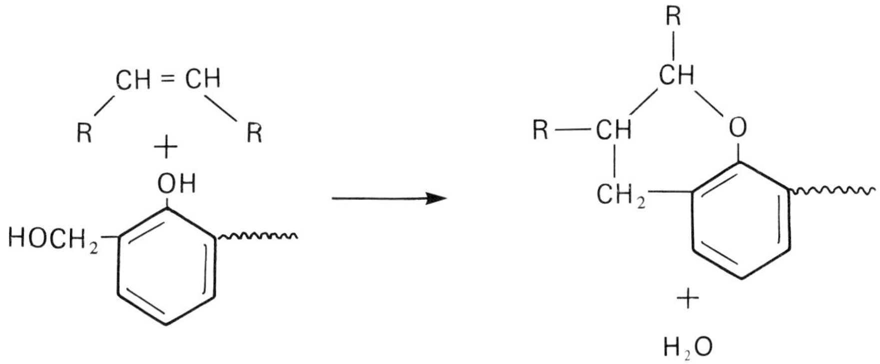

The mechanism is believed to occur (D. H. Solomon, Chemistry of Organic Film Formes, Kreiger Publishing Company, 1977) through quinone methide formation. The ortho and para substituted resoles behave differently.

ORTHO RESOLE

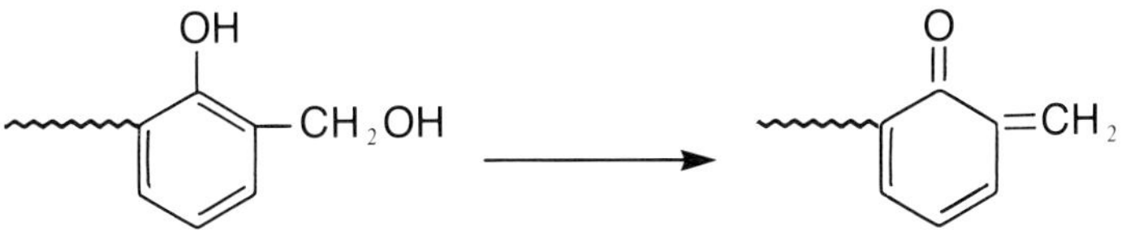

PARA RESOLE

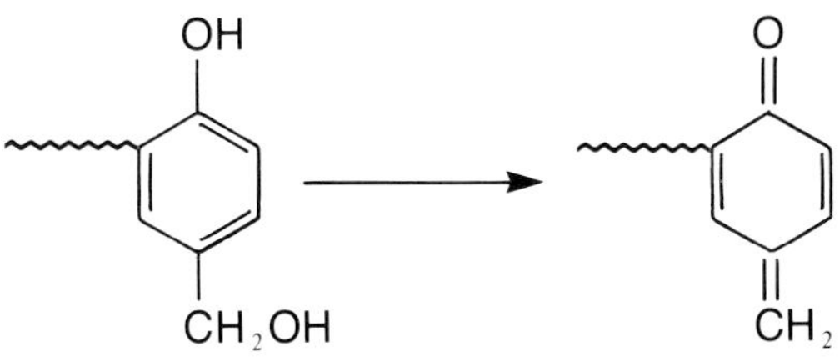

The para resole is sterically hindered for ring formation and reacts with α-methylene groups unlike ortho resoles.

The ortho resoles reaction results in chroman ring formation and is used to introduce phenolic modification into alkyd resins (via the unsaturation of the fatty acid) and gum rosin or rosin esters (via the unsaturation of the abietic acid). Rosin esters treated in this way are utilised in the manufacture of printing inks. (See the Sections on modified alkyd resins and oleoresinous varnishes.)

The probable structure of phenolic modified rosin is:

Rosin (abietic acid)

PHENOL-FORMALDEHYDE NOVOLACS

A novolac is a non-reactive phenolic resin. It is manufactured by reacting a phenol and formaldehyde at an F:P ratio of 1:1, or less, under acidic conditions. Most commercial formulations employ a F:P ratio in the range 0.75−0.85:1. These resins are stable, thermoplastic and soluble in hydrocarbon if made from t-butyl phenol, etc.

A typical novolac structure is shown below.

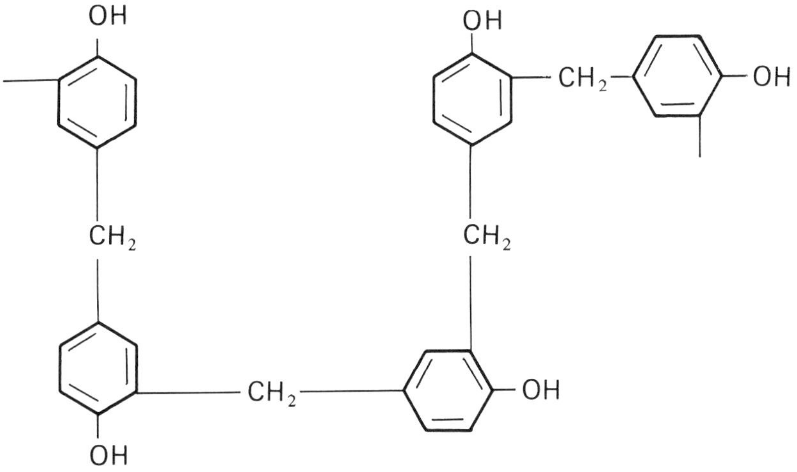

Unlike resoles, novolacs do not contain residual methylol groups. For a novolac to cross link it requires the addition of formaldehyde. In the vast majority of commercial systems formaldehyde is added in the form of hexamine (hexamethylene tetramine).

Hexamine dissociates on heating to give formaldehyde and ammonia. The ammonia which is liberated assists in the development of three-dimensional structures by forming hydroxy benzylamine type structures.

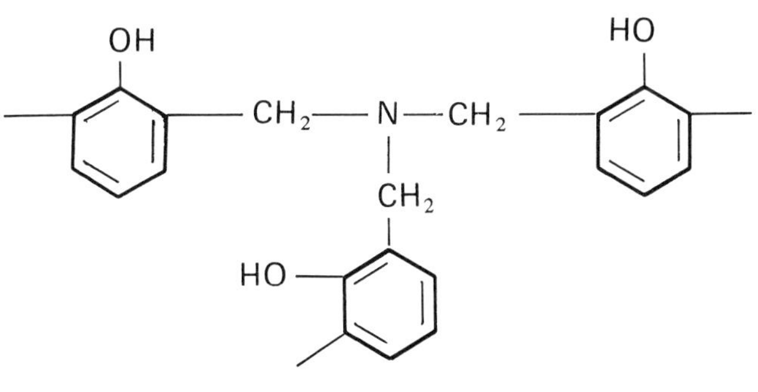

GENERAL COMMENTS ON THE
CURING OF PHENOLIC RESINS

Phenolic resins tend to form brittle films, when used on their own. For this reason they are often used in conjunction with other surface coating resins. Often this combination takes the form of a co-cure system, but phenolics can also be incorporated directly into other resins during manufacture where they provide sites for cross linking and film forming.

An example of incorporation is the modification of alkyd or rosin esters via chroman ring formation. Novolacs are often incorporated into epoxy resins and the resulting epoxy-novolac coatings have excellent adhesion, film strength and flexibility.

The co-cure reactions between phenolic methylol groups and other resins are similar to those detailed under the section dealing with amino resins and will not be reproduced here.

However, there is one important reaction which phenolic resins will undergo which is not found with amino resins. The reaction is the etherification of the phenolic OH group. Both novolacs and resoles undergo this reaction. Etherification of the phenolic OH group leads to improved alkali resistance, better flexibility and enhanced air drying properties, in combination with alkyl compounds.

Alkylation of the OH group is usually brought about by the reaction with strong electrophiles, prior to reaction with formaldehyde.

Typical reactants are alkyl chloride, epichlorohydrin, epoxy resins and monochloroacetic acid.

Allyl prepolymers produced by alkylation with allyl chloride are used as additives for can and drum coatings and for electrodeposition paints.

RAW MATERIALS USED IN COMMERCIAL
MANUFACTURE OF PHENOLIC RESINS

A variety of phenols are used commercially in reactions with formaldehyde. The structure of the resole formed depends on the functionality of the phenol (i.e. number

Phenol Functionality	1	2	3	4
	2,6-Xylenol	o-cresol	Phenol	Bisphenol A
	2,4-Xylenol	p-cresol	m-cresol	
		p-tertiary butyl phenol	3,5-Xylenol	
		p-tertiary butyl cresol	Resorcinol	
		p-nonyl phenol		
		3,4-Xylenol		
		2,5-Xylenol		

of sites available for methylolation). Examples of the most commonly used phenols together with their functionality are given in the table on page 38. Formaldehyde has an overall functionality of two in reaction with phenols.

Formaldehyde

Although not the only carbonyl species used to produce phenolic resins, formaldehyde is by far the most commercially important, and easily the most reactive towards phenols.

Formaldehyde, used in phenolic resin manufacture, is introduced either as an aqueous solution (Formalin) containing approximately 37% by weight of formaldehyde, or in the form of a fine powder or prills (paraformaldehyde).

Formalin

In aqueous medium a very fast acid or based catalysed hydration occurs to form methylene glycol.

$$H - \underset{\underset{O}{\|}}{C} - H \quad + \quad H_2O \quad \longrightarrow \quad HO - CH_2 - OH$$

Mono methylene glycol polymerises to form long chain polymers containing up to 10 repeat units.

$$n\ HO - CH_2 - OH \quad \rightarrow \quad HO(CH_2O)_n - H$$
$$\text{where n lies between 1 and 10}$$

The concentration of non-hydrated formaldehyde in formalin is very low (1%). A 37% aqueous solution of formaldehyde exists as a solution of polymethylene glycol with a molecular weight distribution from $n=1$ to $n=10$. Only in very dilute solutions, below 2%, does the methylene glycol exist as a monomer.

Depolymerisation of aqueous polymethylene glycol occurs in the presence of acid or base catalysts, releasing monomeric methylene glycol for methylolation. The rate of this depolymerisation is important in determining the overall rate of the methylolation reaction.

Paraformaldehyde

Paraformaldehyde is in the form of a powder or prills. It is available commercially in grades of purity from 75% to 97%. A small quantity (usually 1–2%) of methanol is present to act as a stabiliser.

On dissolving paraformaldehyde in water, methylene glycol is generated, as with formalin.

In general paraformaldehyde is not used to the same extent as formalin in the commercial manufacture of phenolic resins. Formalin is usually less expensive and the presence of large amounts of water acts as a heat sink to help dissipate the exothermic heat of reaction. However, during condensation the water from the formalin has to be removed as well as the water produced by the reaction. The use of paraformaldehyde will result in better reactor yields and easier handling. This has to be offset against the higher material costs.

THE MANUFACTURE OF PHENOLIC RESINS

The reaction of phenols with formaldehyde can be potentially dangerous.

It is essential, during manufacture, to exercise rigid control over reaction conditions, particularly temperature and pH. If not properly controlled premature polymerisation can occur leading to a 'run away' exothermic reaction. The consequences of a 'run away' reaction can be serious. Explosions which have destroyed reactor and reactor building have occurred. When using formalin the water present acts as a heat sink, removing exothermic heat of reaction and so helping to moderate the reaction.

Resole Manufacture

F:P ratios between 1:1 and 3:1 are normally employed. Methylolation is carried out under alkaline conditions and the type of catalyst employed has a great influence on the molecular structure and molecular weight distribution.

Catalysts used commercially include sodium hydroxide, alkaline earth oxides and hydroxides, sodium carbonate, ammonia and a range of tertiary amines.

Choice of catalyst is influenced by cost and catalyst performance as well as the ease of separation of the catalyst.

Sodium catalysts are not usually separated from the product, but alkaline earth oxides and hydroxides are usually precipitated as the sulphate and filtered off. The presence of catalyst residues can lead to inferior product performance but this is not so significant when the phenolic resin is intended for use in surface coating applications.

Following methylolation the pH of the reactants is adjusted to below seven and condensation is carried out, usually under vacuum distillation to aid water removal, until a resin with the desired physical properties is achieved, e.g. solids content, viscosity, cure time or melting point depending on the resin type and control testing method.

Novolac Manufacture

Novolacs are produced under acidic pH conditions. The most widely used catalyst being oxalic acid which also acts to give low colour products.

F:P ratios in the range 0.75–0.85 are normally used commercially.

Formaldehyde and phenol are mixed together with catalyst and heated to 95°C. The mixture is held at this temperature for a fixed time period usually about an hour and then water and unreacted phenol are removed under vacuum distillation. Temperatures up to 160°C are often used and sometimes steam sparging is employed to aid removal of unreacted phenol. This stage of the reaction is monitored by melting point determination and as soon as the desired melting point is reached the product is discharged. Residual water has a pronounced effect on the melting point of a novolac, 1% residual water reducing the melting point by 3−4°C. Melt viscosity is also dramatically effected by residual water content. A 90% reduction in melt viscosity at 120°C has been recorded in the literature by increasing the residual water content of a novolac from 0.1% to 3.1%.

The reactivity of novolacs with hexamine increases with increasing free phenol content.

FORMULATIONS

A typical formulation for a resole resin is:

Formaldehyde (37% aqueous solution)	51.85
Sodium hydroxide (30% aqueous solution)	as required
p-tertiary butyl phenol	48.05
Sulphuric acid (50%)	as required
Oxalic acid	0.1
	100.00

PROCESS

1. Charge reactor with formaldehyde and add sodium hydroxide solution to adjust the pH to 8.3−8.5.
2. Add p-tertiary butyl phenol and heat under total reflux conditions to 70°C.
3. Hold at 70°C monitoring the pH, which must not be allowed to fall below 8.3, and the unreacted formaldehyde content.
4. When the unreacted formaldehyde content is indicative of a reacted F:P ratio of 1.8:1 cool the reactor to below 40°C.
5. At 40°C add sulphuric acid (50%) to adjust the pH to 6.5.
6. Stop the stirrer and allow the resin and water to separate into two layers. Remove the aqueous layer by syphon. Add further water, stir and again allow the reactants to separate and remove the aqueous layer. Continue the washing process until a sample of wash water is free of sulphate ions.
7. Add oxalic acid and adjust the pH to 6.0−6.5. Set the reactor for distillation and heat to distillation temperature.
8. Apply 25 inch vacuum and hold under vacuum distillation. The reaction is monitored

by melting point determination. When a sample ex-reactor has a melting point of 140°C the resin is discharged hot into trays.

TYPICAL FINAL PRODUCT CONSTANTS

Appearance	pale amber solid
Melting point	140 – 150°C
Reacted F:P ratio	1.8:1
(Starting F:P ratio)	2.0:1

A formulation for a typical Novolac resin is:

Phenol	20.85
Formaldehyde (37% aqueous solution)	14.82
Oxalic acid	0.16
Water	64.17
	100.00

PROCESS

1. Charge formaldehyde and phenol.
2. Heat to reflux temperature and hold 30 minutes to dissolve the phenol.
3. Add oxalic acid and hold under total reflux conditions at 90–95°C for one hour.
4. Add water and cool the resin to below 50°C.
5. Stop the stirrer and allow reactants to separate into resinous and aqueous layers.
6. Remove the aqueous layer by syphon.
7. Set reactor for vacuum distillation and heat under 25 inch vacuum to 120°C removing water as distillate on the heat up.
8. Hold at 120°C until a sample of resin ex-reactor is brittle at room temperature. Continue to hold at 120°C until a melting point of 60°C is obtained.
9. Discharge the resin hot into trays.

TYPICAL FINAL PRODUCT CONSTANTS

Appearance	pale amber solid
Melting point	60 – 80°C
F:P ratio	0.83:1

PROPERTIES AND APPLICATIONS OF
PHENOLIC RESINS IN SURFACE COATINGS

On their own, phenolic resin films are too brittle to be of any commercial use in surface coating systems. However, in combination with other resins, e.g. alkyd resins, epoxy resins, etc., they form excellent surface coatings giving high adhesion and good chemical resistance. The colour and colour retention of coatings containing phenolic resins is usually poor and so precludes their use in topcoat systems. However, they are widely used as primers and undercoat systems and phenolic modified oil soluble resins are of great importance in printing ink systems.

Automotive primers are an important use area for epoxy phenolic coatings (particularly in electrodeposition systems). However, the dark colour precludes their use as top coats, acrylic-MF or alkyd-MF combinations being preferred for this application.

Phenolic epoxy coatings offer excellent corrosion resistance, sulphur staining resistance and they are frequently used as interior and exterior coatings for drums and cans. Epoxy phenols are extensively used in coatings for food cans.

Modified phenolic resins are frequently used in antifouling marine paints in combination with chlorinated rubber coating systems. Oil or alkyd resins modified with between 25−100% (based on weight of oil) of resole or novolac resins form the basis of anti-corrosive undercoats for both sea and land transport.

Phenolic resins can be rendered oil soluble by incorporation into alkyd and rosin ester resins. Such modified phenolic resoles are of great importance as binders for gravure and offset printing inks, where drying rate is decisive in determining printing press speeds.

FORMULATIONS

EPOXY-PHENOLIC CAN COATING VARNISH

Epoxy resin	25.6
Phenolic resole (F:P ratio 2:1)	14.4
Silicone levelling agent	0.3
Ethylglycol	20.0
Ethylglycol acetate	39.5
Phosphoric acid	0.2
	100.0

Cure schedule: 180°C for 15 minutes

PHENOLIC WASH PRIMER

Phenolic resole	12.7
Polyvinyl butyral	12.1
Epoxy resin	0.9
Zinc tetroxychromate	8.6
Micronised talc	5.7
Aerosil	0.3
Isopropanol	42.9
Wetting agent	0.4
n-Butanol	16.4
	100.0

The above is mixed before application with the following hardener combination:

Phosphoric Acid (85%)	6.0
Isopropanol	29.0

BUTYLATED PHENOLIC RESIN

The straight phenolic resins suffer from brittleness and poor solubility in aromatic solvents. This poor solubility is overcome by the etherification of the methylol groups, butyl alcohol being the most common alcohol used.

Butylation or etherification is usually carried out during the acid stage of the process by refluxing with excess alcohol and removing the water formed by distillation under vacuum or azeotropic conditions.

The resultant resins give films which are hard but flexible with heat and chemical resistance.

A simplified reaction may be represented thus:

FORMULATION

A typical formulation for a butylated phenolic resin is:

Phenol	23.0%
Paraformaldehyde	17.0%
Ammonia (35%)	1.4%
Phosphoric acid	0.9%
Butyl alcohol	57.7%

PROCESS

1. Charge reactor with phenol and paraformaldehyde and quarter of the ammonia.
2. Heat slowly to 50°C and hold for one hour.
3. Add a further quarter of the ammonia and increase the temperature to 50°C and hold for a further hour.
4. Add another quarter portion of ammonia, increase temperature to 70−95°C and hold for one and a half hours.
5. Add the final portion of ammonia and hold until a viscosity of 1.5−2.0 poise is obtained for a 60% solution in butanol.
6. Cool and add the phosphoric acid and 80% of the butanol.
7. Raise temperature and remove water azeotropically until temperature is 116°C.
8. Add the remainder of the butanol and distil off under atmospheric conditions until the non-volatile content is 45% and the temperature is 118°C.

TYPICAL FINAL PRODUCT CONSTANTS

Appearance	pale amber liquid
Non-volatile content	45%
Viscosity	1.5 poise
(Starting F : P ratio)	2.3 : 1

A considerable improvement in the flexibility and resistance properties may be obtained by blending a phenolic resin, similar to that used above, with epoxy resins having E.E.W. in the range 1700−4000.

Ratio of epoxy to phenolic should be between 1 : 1 and 1 : 2, these may be cold blended together or co-condensed at 70°C for one hour.

AMINO RESINS

The term amino resin is applied to resinous products formed from the reaction between formaldehyde and compounds containing amino groups, i.e. amide, amine or imide groups.

Amino resins are widely used in surface coating systems normally in combination with other polymeric species.

The most commonly used amino resins are those produced from urea, melamine and to a lesser extent benzoguanamine.

Vinyl amides such as acrylamide or methacrylamide are also reacted with formaldehyde, after copolymerisation with other vinyl monomers, to provide sites for curing of acrylic polymers.

AMINO RESIN-FORMALDEHYDE REACTIONS

Amide-Formaldehyde Reaction

Consider the simple amide:

$$R - \underset{\underset{O}{\|}}{C} - NH_2$$

Under the conditions existing during manufacture only one of the H atoms is available for reaction:

$$\underset{\substack{\| \quad | \\ O \quad H}}{R - C - NH} \quad + \quad HCHO \quad \longrightarrow \quad \underset{\substack{\| \quad | \\ O \quad CH_2OH}}{R - C - NH}$$

Methylol amide

The methylolation of a simple amide will not lead to polymer formation on condensation:

$$\underset{\substack{\| \quad | \\ O \quad CH_2OH}}{R - C - NH} \quad + \quad \underset{\substack{| \quad \| \\ HOCH_2 \quad O}}{HN \underline{\quad\quad} C - R}$$

$$\downarrow$$

$$\underset{\substack{\| \quad | \\ O \quad CH_2 - O - CH_2}}{R \underline{\quad} C \underline{\quad} NH} \quad \underset{\substack{| \quad \| \\ \quad\quad O}}{HN \underline{\quad\quad} C \underline{\quad} R}$$

In order to form a polymer a diamide is required. Simple monofunctional amides are sometimes added as reaction modifiers in the polymerisation of a multi-functional amide.

Urea-Formaldehyde reaction

Although strictly an amide, urea, when reacted with formaldehyde behaves as though it had an amine and an imide functionality.

$$\underset{\substack{\| \\ O}}{H_2N - C - NH_2} \quad \rightleftharpoons \quad \underset{\substack{| \\ OH}}{H_2N - C = NH}$$

(urea)   (amine)   (imide)

Because of this only three hydrogen atoms are available for replacement.

METHYLOLATION

The first step in the urea formaldehyde reaction is the formation of methylol groups at the sites of the labile H atoms.

$$H_2N - C - NH_2 \quad + \quad HCH \quad \rightleftharpoons \quad H_2N - C - NH$$
$$\overset{\|}{O} \qquad\qquad \overset{\|}{O} \qquad\qquad \overset{\|}{O} \quad \overset{|}{CH_2OH}$$

Monomethylol urea

$$H_2N - C - NH \qquad\qquad H - C - H \quad \rightleftharpoons \quad HN - C - NH$$
$$\overset{\|}{O} \ \overset{|}{CH_2OH} \quad + \quad \overset{\|}{O} \qquad\qquad \overset{|}{HOH_2C} \ \overset{\|}{O} \ \overset{|}{CH_2OH}$$

Dimethylol urea

Although trimethylol urea may be formed, under the conditions that normally pertain during commercial manufacture, only di- and mono-methylol ureas are formed. The number of methylol groups formed depend on the mole ratio of formaldehyde to urea, and the pH and temperature under which the reaction is carried out. In general a high pH favours the methylolation reaction.

POLYMERISATION

Once formed the methylol groups react to give a polymerised structure by loss of water and/or formaldehyde.

i) $\sim\sim\sim NHCH_2OH \quad + \quad NH\sim\sim\sim \quad \rightarrow \quad \sim\sim\sim NHCH_2 - N\sim\sim\sim$
$$\overset{|}{CH_2OH} \qquad\qquad\qquad \overset{|}{CH_2OH} \quad +H_2O$$

ii) $\sim\sim\sim NHCH_2OH \quad + \quad HOCH_2NH\sim\sim\sim \quad \rightarrow \quad \sim\sim\sim NHCH_2OCH_2NH\sim\sim\sim$
$$+ H_2O$$

iii) $\sim\sim\sim NHCH_2OH \quad + \quad HOCH_2NH\sim\sim\sim \quad \rightarrow \quad \sim\sim\sim NHCH_2NH\sim\sim\sim$
$$+ HCHO + H_2O$$

iv) 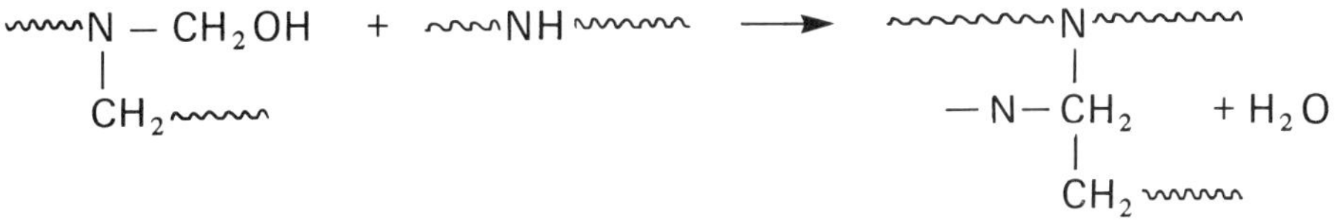 $\sim\!\!\sim\!\!N - CH_2OH \quad + \quad \sim\!\!\sim\!\!NH\sim\!\!\sim\!\!\sim \quad \longrightarrow$

with products:

$\sim\!\!\sim\!\!\sim\!\!\sim\!\!N\sim\!\!\sim\!\!\sim\!\!\sim$

$-N-CH_2 \quad + H_2O$

$CH_2 \sim\!\!\sim\!\!\sim$

(on left reactant: N bonded to $CH_2\sim\!\!\sim$)

Complicated structures are formed by these reactions and a typical polymerised product was thought to have a trimer structure similar to that shown below and suggested by Marvel.

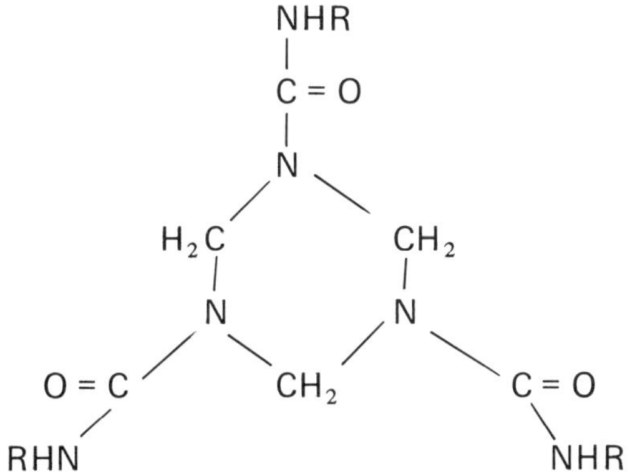

Where R may be CH_2OH or a H atom.

This theory is no longer supported and modern evidence indicates the presence of molecular chains during the early stage of condensation.

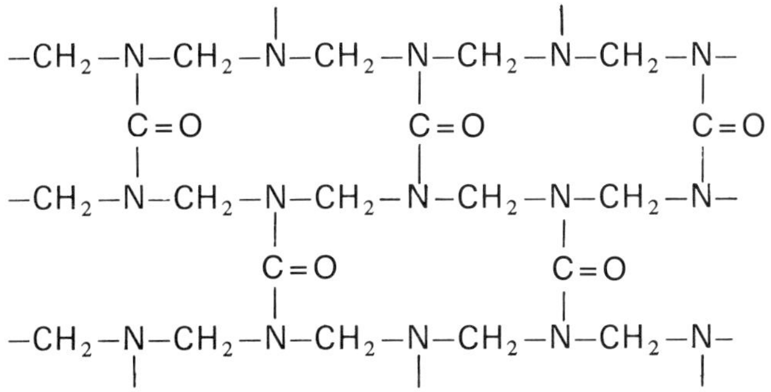

The reaction between the amino groups and formaldehyde is a very complex one consisting of several simultaneous ones. Further heating will 'cure' the resin forming a rigid cross linked structure as shown below. This differs from the idealised structure having lost formaldehyde and containing residual CH_2OH and NH groups.

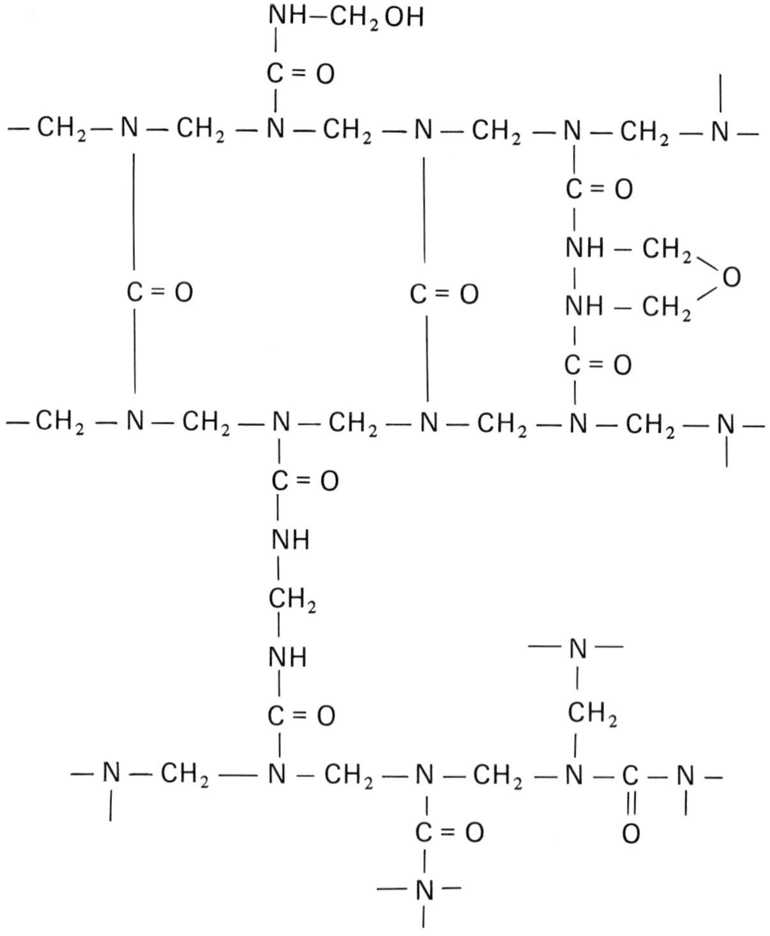

Cross linked urea-formaldehyde resin

The polymeric product is insoluble in most common organic solvents and has a short shelf life. It readily cures on application of heat or on treatment with an acidic catalyst to form an infusable, insoluble cross linked material.

This type of resin is used for chipboard bonding, moulding compositions and adhesive applications, and is unsuitable for use in surface coating systems, due to its poor stability, solvent tolerance and incompatibility with other resins.

To achieve a product suitable for use as a surface coating it is necessary to restrict the amount of condensation occurring due to the methylol-methylol reaction by removing methylol groups. This is carried out by etherification of some of the methylol groups with alcohol to form alkyl ethers.

The degree of alkylation or etherification can be controlled by the temperature, reaction time, pH and quantity of alcohol used. In general, low pH with excess alcohol favours alkylation over the competing polymerisation reaction. Usually the alcohols used for etherification are n-butanol, iso-butanol ethanol or methanol.

ETHERIFICATION WITH ALCOHOL (alkylation)

Amino-Formaldehyde Reaction

The only other amines whose reactions with formaldehyde are of commercial importance are melamine and benzoguanamine.

Melamine   Benzoguanamine

The reaction route for both compounds is similar to that for urea, i.e. methylolation followed by polymerisation and alkylation.

The number of replaceable hydrogen atoms for melamine is six and for benzoguan-amine it is four. This can lead to the rapid formation of a more cross linked structure than when urea is used, and hence rigid control of the initial methylolation and degree of alkylation is essential when producing useful amino resins from these amines.

A major difference between melamine and urea in its reactions with formaldehyde are the conditions under which methylolation occurs. Unlike urea, melamine readily undergoes both methylolation and polymerisation under acidic conditions with comparable reaction rates. The ratio of melamine to formaldehyde is a major factor in the influencing of the polymerisation rate. At high F:M ratios (i.e. 6:1) polymerisation is generally slower than at a F:M of 3:1. In general various methylol melamines are formed:

Mono   Tri   Hexa

It is considered that under acidic conditions methylol melamines undergo condensation reactions and both straight or branched chains of triazine units linked with methylene ether and methylene groups are built up.

$$2 \text{ RNHCH}_2\text{OH} \xrightarrow[\text{conditions}]{\text{Acidic}} \text{RNHCH}_2-\text{O}-\text{CH}_2 \text{ NHR} + \text{H}_2\text{O}$$

The properties of the product depend upon similar factors as for urea formaldehyde resins and are those which affect the relative degrees of polymerisation and alkylation. These are:

a) The molar proportions of reactants, formaldehyde, melamine, alcohol and ratio of alcohol to water.

b) The reaction media.

c) Type of catalysts used and pH values during the reaction.

d) Temperature and duration of reaction, condensation and/or etherification.

A typical structure for a butylated melamine-formaldehyde resin is:

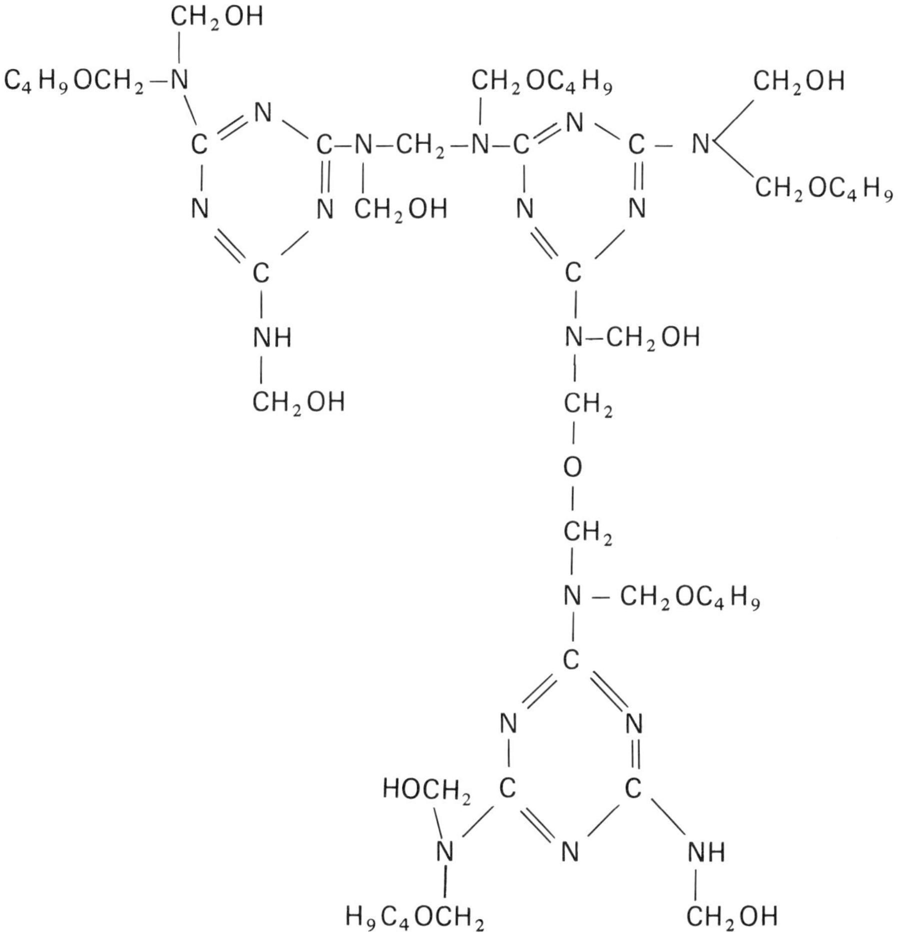

$$
\begin{array}{c}
\text{CH}_2\text{OH} \\
| \\
\text{C}_4\text{H}_9\text{OCH}_2-\text{N}
\end{array}
$$

CH₂OH
|
C₄H₉OCH₂—N
C=N
C—N—CH₂—N—C=N
C — N
CH₂OC₄H₉

CH₂OC₄H₉ CH₂OH

N N CH₂OH N N
C C
NH N—CH₂OH
CH₂OH CH₂
O
CH₂
N — CH₂OC₄H₉
C
N N
HOCH₂ C C
N N NH
H₉C₄OCH₂ CH₂OH

The only aldehyde used in any quantity for the manufacture of amino resins is formaldehyde. It is, however, available in three forms:

Formaldehyde (aqueous solution)

Paraformaldehyde (formaldehyde concentrates)

Solutions of formaldehyde in alcohol.

Formaldehyde in aqueous solution consists mainly of formaldehyde gas in hydrated form. It exists as an equilibrium mixture of monohydrate methylene glycol, $CH_2(OH)_2$ with low molecular weight polymeric hydrates $HO(CHO_2O)_nH$. Usually the solutions contain

36–37% formaldehyde, 1–10% methanol (depending upon the grade), with the remainder being water.

Paraformaldehyde is a mixture of polyoxymethylene glycols having the generic formulation $HO(CH_2O)_nH$ and contains 8–100 formaldehyde units per molecule. Formaldehyde contents vary between 75–97% with 1–5% methanol and the remainder being water. They are supplied as flakes, fine powder or prills.

Formaldehyde solutions in alcohols are equilibrium mixtures of hemiacetals of the particular alcohol and formaldehyde. They are available from butanol, isobutanol or methanol as the alcohol. Their composition varies between 40–50% formaldehyde and 45–55% alcohol, with the remainder being water.

All three types are used in amino production although paraformaldehydes have the advantages of higher yields, and faster reaction times with less water being removed in the distillation process. Alcohol solutions are easier to handle. The paraformaldehyde does not have to be solubilised before commencement of the process. They contain minimal amounts of water and contain both essential reactants. Processes are simpler with faster process times.

Melamine resins are becoming of increasing importance as cross linking agents for water dilutable surface coatings, particularly in the textile and adhesive sector. Resins suitable for use in this application are low molecular weight, highly alkylated resins. The resin is prepared by fully methylolating melamine to form hexamethylol melamine and then fully alkylating the methylol groups with low molecular weight alcohols such as methanol.

Factors influencing the Choice of Alcohol used for Alkylation

The alcohol used to etherify the methylol groups must fulfil the following requirements.

   i) It must form an azeotrope with water that will separate on cooling. This is essential so that water of reaction can be efficiently removed from the reactor without decreasing the amount of alcohol available for reaction. Methanol is an exception and usually involves different processes.

   ii) It must form an alkyl ether that will readily interact with other polymeric species during co-cure.

   iii) It must react readily with methylol groups in the presence of the competing polymerisation reaction.

   iv) It must be a solvent for the methylolated amino compound.

The most widely used alcohols are 'normal' and 'iso butanol' although other alcohols may be used where specific solubility or performance properties are required from the product.

In general, primary alcohols form alkyl ethers much faster than secondary or tertiary alcohols.

The longer the chain length of the alcohol the slower the alkylation reaction, but the greater the tolerance of the product for hydrocarbon solvents.

Because of their relatively slow reaction rate, the amount of alkylation is small compared to the amount of polymerisation when the higher alcohols are reacted with methylolated amino compounds. Thus, when long chain alcohols are required for alkylation, it is common practice first to form butyl ether and then use an ester inter-change reaction to exchange the butanol for a longer chain alcohol.

Ester Inter-Change Reaction

$$\text{\textasciitilde\textasciitilde\textasciitilde NH} \quad + \quad R\text{–OH} \quad \longrightarrow \quad \text{\textasciitilde\textasciitilde\textasciitilde NH}$$
$$| \qquad\qquad\qquad\qquad\qquad\qquad\qquad\qquad |$$
$$CH_2OC_4H_9 \qquad\qquad\qquad\qquad\qquad\qquad CH_2OR$$

$$+$$

$$C_4H_9OH$$

where R = C_5 or higher alkyl group

As the chain length of the alcohol used for alkylation increases, so the rate at which cure occurs decreases.

The use of unsaturated alcohols (e.g. allyl alcohol) for alkylation results in an amino resin with a potential for oxidation cure. This technique can be used successfully where an otherwise slow curing resin gives rise to performance deficiencies.

Effect of the Degree of Alkylation

The degree of alkylation is a measure of the number of methylol groups that are etherified and is an important factor in determining the performance characteristics of the amino resin.

As the degree of alkylation increases, the viscosity and reactivity of the resin decreases, while compatibility with other surface coating resins increases. During manufacture care must be exercised to achieve the desired balance between the degrees of methylolation, alkylation and polymerisation which occur.

THE CURING OF AMINO RESINS

On heating at temperatures of 120–150°C an alkylated amino-formaldehyde will cure to form a clear, coherent film by reaction between adjacent molecules. Four separate reactions can occur, dependent on the nature of the reacting group.

i)

$$\text{\textasciitilde\textasciitilde –NH–\textasciitilde\textasciitilde} + \text{\textasciitilde\textasciitilde N\textasciitilde\textasciitilde} \longrightarrow$$

$$CH_2OC_4H_9 \qquad\qquad \text{\textasciitilde\textasciitilde N\textasciitilde\textasciitilde}$$
$$| \qquad\qquad\qquad\qquad\qquad |$$
$$\qquad\qquad\qquad\qquad\qquad CH_2 \qquad + C_4H_9OH$$
$$\qquad\qquad\qquad\qquad\qquad |$$
$$\qquad\qquad\qquad\qquad\text{\textasciitilde\textasciitilde N\textasciitilde\textasciitilde}$$

ii)

$$\underset{\underset{\displaystyle \mathsf{N-CH_2OH}}{\overset{\displaystyle |}{}}}{\overset{\displaystyle CH_2OH}{}} + HOCH_2N\text{\textreferencemark} \longrightarrow \text{\textreferencemark}N-CH_2-N\text{\textreferencemark} + HCHO + H_2O$$

CH₂OH / ⁓⁓N – CH₂OH + HOCH₂N⁓⁓ → ⁓⁓N – CH₂ – N⁓⁓ (CH₂OH, CH₂OH substituents) + HCHO + H₂O

iii)

CH₂OH / ⁓⁓N – CH₂OC₄H₉ + HOCH₂N⁓⁓ → ⁓⁓N–CH₂–N⁓⁓ (CH₂OC₄H₉, CH₂OH, CH₂OC₄H₉ substituents) + HCHO + C₄H₉OH

iv)

CH₂OC₄H₉ / ⁓⁓N–CH₂OH + HOCH₂–N⁓⁓ → ⁓⁓N–CH₂–O–CH₂–N⁓⁓ (CH₂OC₄H₉, CH₂OC₄H₉, CH₂OC₄H₉ substituents) + H₂O

The films produced from alkylated amino resins on stoving, are not sufficiently flexible for most surface coating applications and their adhesion, particularly on metal substrates, is poor. Amino resins are therefore used in combination with other resins such as epoxy resins, alkyds and hydroxy functional acrylic resins.

The combinations of amino resins with other resin species produce surface coating systems combining the beneficial properties of both resin types. For example an alkyd resin/melamine resin combination produces a cured film combining the flexibility and adhesion properties of the alkyd with the chemical resistance and hardness properties of the melamine resin.

GENERAL COMMENTS ON CO-CURE OF AMINO RESINS

Amino resins are usually classed as slow or fast curing. The cure rate refers to the hardness developed on film forming. In general, the faster cure resins result in less co-reaction between polymer species. This comes about because the methylol-methylol reaction is faster than the methylol-hydroxyl cross linking reaction. Thus a high proportion of the inter polymer species links are formed by the slower reacting alkylated methylol groups.

This is often reflected in the long term exposure performance of the coatings and it is often the case that the hardest film results in the poorest weathering performance.

The cure rate of urea-formaldehyde resins can be improved by the use of acidic catalysts. In general the greater the amount of acid catalyst present the faster the cure.

Melamine resins do not, in general, have the same speed of cure when acid catalysed, particularly when cured at room temperature. They are however sensitive and, in the presence of acid catalysts, the cure rate will increase with the level of acid until a maximum is reached.

Increasing the level of catalyst above this point will not result in an increased cure rate and often leads to adverse film performance and poor stability.

Amino Resin — Alkyd Resin Co-Cure Systems

Amino resins are frequently used in conjunction with alkyd resins based on non-drying oils such as castor oil, coconut oil, etc. The alkyd resin is formulated to contain free hydroxyl groups and residual acid value. The OH groups act as sites for cross linking while the acid groups act as a catalyst for the cure reaction.

CURE REACTIONS

THE PROPERTIES OF ALKYD-AMINO RESIN COATING SYSTEMS

In general increasing amino resin content leads to increase in film hardness, colour retention properties and chemical resistance. Increasing alkyd resin content leads to an increase in film flexibility and adhesion properties.

Although the cross linking reactions with an alkyd resin are similar for both urea and melamine formaldehyde resins there are pronounced differences in the performance properties of films produced from urea-alkyd and melamine-alkyd systems.

Advantages of Melamine-Alkyd Combinations

i) Melamine-alkyd systems have faster cure times than a corresponding urea-alkyd system and produce harder films.

ii) Stoving temperatures up to 250°C may be used for melamine-alkyd systems while temperatures above 150°C are not recommended for urea-alkyd systems, as the urea resin breaks down.

iii) Melamine-alkyd systems generally result in better gloss and gloss retention.

iv) Melamine-alkyd systems give better colour retention on overstove.

v) Melamine-alkyd systems have superior chemical resistance and exterior durability.

The Advantages of Urea-Alkyd Combinations

i) Urea-alkyd systems have better adhesion particularly with metal substrates. When using melamine-alkyd systems on metal it is necessary to prime the surface well before application. Alternatively 20−30% of the melamine resin of the melamine-alkyd combination may be replaced by a urea based resin.

ii) Urea based resins are lower cost. However, some of the cost advantage is offset by the fact that a higher proportion of urea resin is required to produce the same film hardness results as a melamine-alkyd system. In general, three parts by weight of melamine resin will produce a similar film hardness result as five parts of urea resin.

iii) Urea-alkyd systems can be used in 'cold-cure' applications since they can be cured at ambient temperatures in the presence of acid accelerators.

THE PROPERTIES OF
AMINO-EPOXY RESIN CO-CURE SYSTEMS

Epoxy Resin−Amino Resin Co-Cure Systems

Complex three-dimensional cross linked structures can be built up by stoving amino resin−epoxy resin systems at temperatures above approximately 150°C.

The cured films combine the best features of epoxy resin and amine resins and have excellent adhesion, chemical resistance and gloss retention properties, colour and colour retention and, with correct choice of amino resin, very good flexibility.

These types of co-cure system find extensive use in metal decorating applications. Melamine resins often form the amino resin component in preference to urea. This results in superior film hardness and gloss properties.

This type of co-cure system is normally used where high performance demands outweigh cost considerations.

CURE REACTIONS

i)

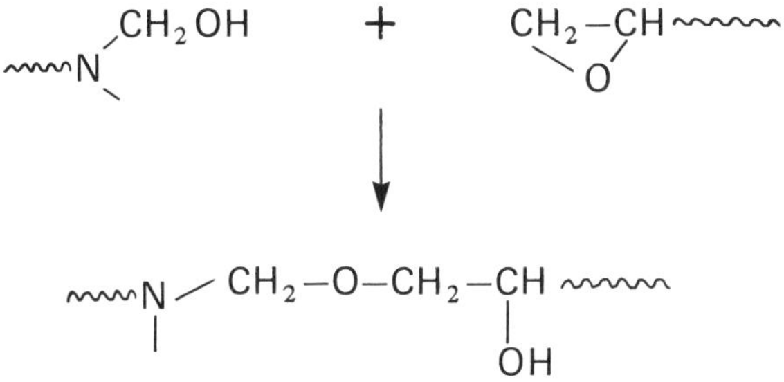

$\sim\sim\sim N \stackrel{CH_2OH}{\diagdown}$ $+$ $CH_2{-}CH\sim\sim\sim$ with O bridge

$\downarrow$

$\sim\sim\sim N{-}CH_2{-}O{-}CH_2{-}CH\sim\sim\sim$
                                    $|$
                                    OH

ii)

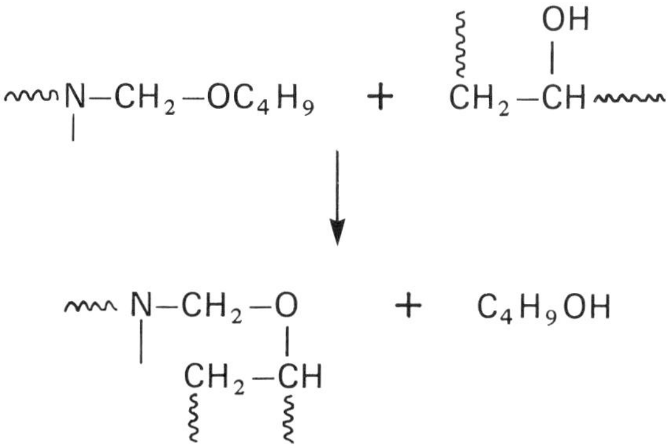

$\sim\sim\sim N{-}CH_2{-}OC_4H_9$ $+$ $\begin{matrix} OH \\ | \\ CH_2{-}CH\sim\sim\sim \end{matrix}$

$\downarrow$

$\sim\sim N{-}CH_2{-}O$ $+$ C_4H_9OH
       $|$ $|$
               $CH_2{-}CH$

iii)

$\sim\sim\sim N{-}CH_2OH$ $+$ $CH_2{-}CH\sim\sim\sim\stackrel{OH}{}\sim\sim\sim CH{-}CH_2$ with epoxide O bridges

$\downarrow$

$\sim\sim\sim N{-}CH_2$ $+$ H_2O
         $|$
         O
$CH_2{-}CH\sim\sim\sim CH{-}CH_2$ with O bridges

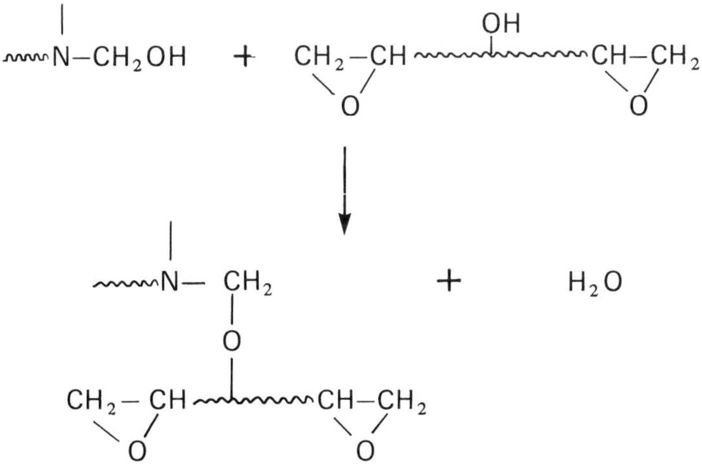

THE PROPERTIES OF
AMINO-ACRYLIC RESIN CO-CURE SYSTEMS

Amino Resin — Acrylic Resin Co-Cure Systems

It is relatively easy to produce acrylic polymers that contain OH groups and/or COOH groups (see chapter on Acrylic resins). Reaction of these resins with amino formaldehyde condensates gives rise to a surface coating which combines the properties of both amino and acrylic resins.

The reaction between amino resins and acid functional groups is relatively slow and considerable self condensation of the amino resin occurs. These systems are usually cured by stoving at temperatures of 150°C for 30 minutes.

Where the acrylic polymer contains hydroxy functional groups the reaction with amino resins is much quicker and stoving temperatures of 125°C are used.

The use of external acid catalysts increases the co-cure rate but often results in poor stability of the coating.

Incorporation of both hydroxyl and carboxyl groups into the acrylic polymer results in a rapid co-curing reaction between amino resin and the hydroxyl groups with the carboxyl groups acting as an internal catalyst.

Amino resin-acrylic systems combine good colour retention with good adhesion and excellent chemical resistance and weathering properties. They are used as the basis for appliance and general industrial enamels. Their excellent polishing and outdoor exposure properties lead to their use in automobile finishes.

CURE REACTIONS

i) With acid functional acrylic polymers

$$\sim\!\sim\!NH-CH_2OH \qquad + \qquad \sim\!\sim\!\boxed{Acrylic\ copolymer}\!\sim\!\sim$$
$$\underset{COOH}{|}$$

$$\downarrow$$

$$\sim\!\sim\!\boxed{Acrylic\ copolymer}\!\sim\!\sim \quad + \qquad H_2O$$
$$\sim\!\sim\!NH-CH_2-O-\overset{|}{C}=O$$

ii) $\sim\!\sim\!NH-CH_2OC_4H_9 \qquad + \qquad \sim\!\sim\!\boxed{Acrylic\ copolymer}\!\sim\!\sim$
$$\underset{COOH}{|}$$

$$\downarrow$$

$$\sim\!\sim\!\boxed{Acrylic\ copolymer}\!\sim\!\sim$$
$$\sim\!\sim\!NH-CH_2-O-\overset{|}{C}=O \quad + \quad C_4H_9OH$$

iii) With hydroxy functional acrylic polymers

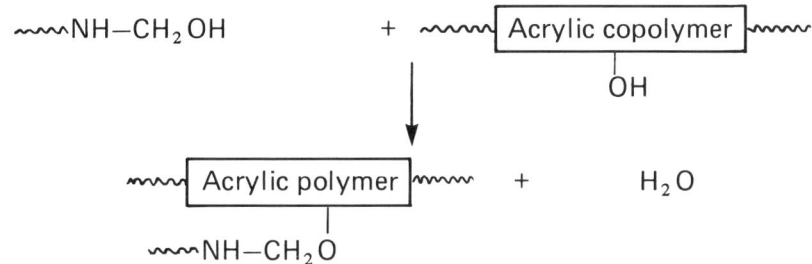

iv)

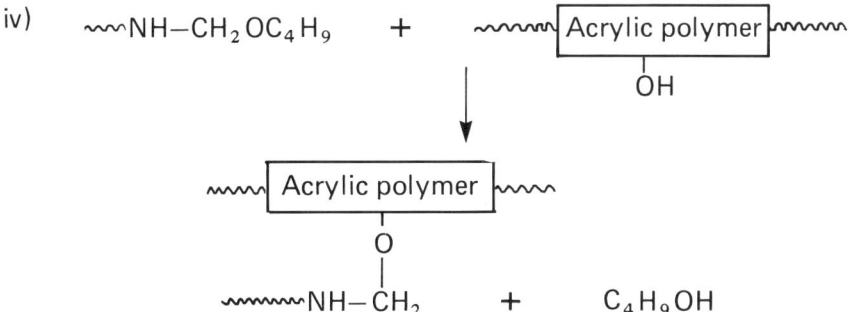

FORMULATIONS

A typical isobutylated urea-formaldehyde resin formulation is:

Isobutanol	50.98
Paraformaldehyde (95%)	30.56
Urea	17.87
Phosphoric acid	0.51
Sodium hydroxide (33% aqueous solution)	0.05
Triethylamine	0.03
	100.00

PROCESS

1. Charge reactor with isobutanol, paraformaldehyde and sodium hydroxide solution.

2. Heat to 60°C to dissolve the paraformaldehyde then charge triethylamine and urea.

3. Heat to 85°C and hold at this temperature for 30 minutes. During this stage frequent checks must be made on the pH of the reactants. The pH must be maintained in the range 8.4−8.6.

4. Cool to 70°C, checking the unreacted formaldehyde content on the cool down. This should be approximately 7% if methylolation has proceeded to the required degree.

5. Add phosphoric acid and adjust the pH into the range 3.5−4.0.

6. Heat to reflux temperature under solvent process conditions. Maintain the temperature at reflux, removing water from the separator and returning the isobutanol to the reactor for four hours, during which time 15.0 parts of water per 100 parts reactor charge should have distilled off. Note, during the distillation stage the reflux temperature will rise to approximately 108°C as water is removed from the reactor.

7. Sample and determine the tolerance of the sample for toluene. When a mixture of one part sample with 10 parts toluene is completely clear at 25°C, the reaction is deemed to be complete.

8. Cool and filter.

TYPICAL RESIN CONSTANTS

Non-volatile content	60%
Brookfield viscosity at 25°C	10 poise
Clarity	very slightly hazy

A typical formulation for a melamine-formaldehyde resin is:

Parformaldehyde (95%)	24.05
n-Butanol	53.54
Melamine	19.20
Phthalic anhydride	0.45
Urea	1.38
Water	1.38
	100.00

PROCESS

1. Charge the reactor with butanol and paraformaldehyde and heat to 60°C to dissolve the paraformaldehyde.

2. Cool to 50°C add melamine and add phthalic anhydride. Adjust pH to 5.9−6.1.

3. Heat to reflux temperature and hold at this temperature removing water from the separator and returning butanol to the reactor. During this stage of manufacture it is essential to maintain the pH in the range 5.9−6.1.

4. Continue to hold under solvent process until no more water is present in the distillate, then cool to below 100°C and add urea pre-dissolved in water.

5. Reheat to reflux temperature and hold under solvent process until the amount of water distillate slows below ½% of the batch weight per hour.

6. Sample and determine the tolerance of the sample for white spirit. When a mixture of one part sample to three parts white spirit is clear at 25°C the process is deemed complete.

7. Cool and filter press.

TYPICAL RESIN CONSTANTS

Non-volatile content	61%
Brookfield viscosity	18 poise
Clarity	clear resin

Formulation for typical melamine formaldehyde resin using formaldehyde in butanol:

Formaldehyde in Butanol*	62.74
n-Butyl alcohol	17.0
Caustic soda	0.02
Melamine	17.8
Xylol	2.4
Phthalic anhydride	0.04
	100.00

*Synthite Limited

PROCESS

1. Charge reactor with formaldehyde in butanol, n-butyl alcohol, melamine and adjust to pH 8−8.5 with caustic soda dissolved in water.

2. Heat to 60°C and hold for one hour.

3. Add xylol and adjust to pH 4−5 with phthalic anhydride.
4. Heat to distillation and continue distilling, removing water azeotropically until temperature reaches 116°C.
5. Sample for white spirit tolerance and viscosity.
6. Either cool if correct, or continue distillation, possibly with the addition of extra butanol until the tolerance is correct.
7. Cool and filter.

TYPICAL RESIN CONSTANTS

Non-volatile content	60%
Viscosity	3 − 4 poise
White spirit tolerance	4 − 5 ml
Initial melamine: Formaldehyde ratio	1 : 6.0

Formulation for urea formaldehyde resin using formalin:

Urea	15.7
Formalin	48.575
Caustic soda	0.025
Phthalic anhydride	0.10
Xylol (1)	2.4
Butanol (1)	24.0
Butanol (2)	5.2
Xylol (2)	4.0
	100.000

PROCESS
1. Charge urea and formalin and adjust to pH 8−8.5 with caustic soda in water.
2. Heat to 70°C and hold for one hour.
3. Vacuum distil at 730−700 mm Hg removing 15% of total charge as water.
4. Cool and add butanol (1), xylol (1) and adjust to pH 4−4.5 with phthalic anhydride.
5. Distil water off azeotropically until temperature reaches 112°C and then add the remainder of the butanol (2) and continue distillation until temperature is 118−120°C.

6. When viscosity and tolerance are within range (viscosity 5−8 poise tolerance 2−3 ml) apply cooling.
7. Determine non-volatile content and thin to 55% non-volatile content with xylol (2) and extra butanol as necessary.

TYPICAL RESIN CONSTANTS

Non-volatile content	55%
Viscosity	4 − 7 poise
White spirit tolerance	2 − 3 ml
Initial urea formaldehyde ratio	1 : 2.3
Butanol: Xylol ratio	1 : 4

COATING FORMULATIONS

The following two formulations are based on the same alkyd resin and illustrate the difference in amino resin content when formulated with urea resins as opposed to melamine resins for a general purpose white stoving enamel.

Urea resin based formulation:

Rutile titanium dioxide	26.50
Castor oil alkyd (50% non-volatile content)	38.90
Butylated urea resin	20.38
Xylol	10.65
n-Butanol	3.57
	100.00

Vehicle solids	45%
Viscosity at 24°C	300 centipoise
Cure schedule	30 mins at 120°C
Alkyd:amino resin ratio (non-volatile content)	60:40

Melamine resin based formulation:

Rutile titanium dioxide	27.60
Castor oil alkyd (50% non-volatile content)	47.90
Butylated melamine resin	17.27
Xylol	5.43
n-Butanol	1.80
	100.00

Vehicle solids	49.0%
Cure schedule	30 mins at 120°C *or* 20 mins at 150°C
Viscosity at 25°C	300 centipoise

Nitro-cellulose modified one-pack low gloss acid hardening lacquer:

Castor oil alkyd (50% non-volatile content)	19.46
Butylated urea resin (60%)	18.00
½ second nitro-cellulose (70% in butyl acetate)	11.09
Ethyl acetate	10.22
Methyl isobutyl ketone	6.94
n-Butanol	17.03
Toluene	10.78
Acid butyl phosphate	1.00
Xylol	0.61
Amorphous silica	4.87
	100.00

Non-volatile content	33%
Vehicle resin solids	28.3%
Viscosity at 25°C	42 secs No. 4 flow cup
Cure schedule	10 mins at 80°C
Alkyd : urea : nitro-cellulose ratio	(1.0 : 1.1 : 0.8)

White enamel for flexible tubes:

40% FA content DCO epoxy ester (58% non-volatile content)	48.31
Rutile titanium dioxide	32.31
Butylated melamine resin (55% non-volatile content)	4.83
Cobalt naphthenate (6% Co)	0.16
Shellsol A	14.49
	100.00

Non-volatile content	63%
Viscosity at 25°C	540 centipoise
Cure schedule	10 mins at 160°C
Pigment : binder ratio	51 : 49
Epoxy ester : melamine ratio	10.6 : 1

Cream venetian blind paint:

Yellow oxide	1.43
Zinc oxide	2.39
Rutile titanium dioxide	23.89
DCO castor oil alkyd (50% non-volatile content)	40.61
Butylated urea resin (50% non-volatile content)	3.82
Butylated melamine resin (50% non-volatile content)	7.64
Xylol	13.76
n-Butanol	6.45
	100.00

Non-volatile content	53.7%
Viscosity at 25°C	400 centipoise
Cure schedule	30 mins at 150°C
Alkyd : urea : melamine ratio	10.6 : 1 : 2

Highly resistant primer − epoxy melamine based:

Anti-settling agent (clay)	0.30
Rutile titanium dioxide	15.09
Zinc oxide	5.03
Zinc chromate	5.03
Micronised talc	10.06
325 mesh Mica	2.52
40% FA content DCO epoxy ester (50% non-volatile content)	31.69
Butylated melamine resin (50% non-volatile content)	8.55
Xylol	21.73
	100.00

Non-volatile content	58%
Cure schedule	120°C for 30 mins
Pigment/binder ratio	1 : 1.3
Epoxy ester : melamine ratio	3.7 : 1

Green gloss general purpose stoving finish — acrylic:

Chromastral green	15.55
Hydroxy functional acrylic (55% non-volatile content)	52.28
Butylated melamine resin (55% non-volatile content)	16.32
Xylol	12.68
n-Butanol	3.17
	100.00

Non-volatile content	53%
Cure schedule	30 mins at 120°C
Acrylic : melamine ratio	3.2 : 1

Yellow drum enamel:

Chrome yellow	28.50
Linseed oil alkyd (70% non-volatile content)	53.20
Melamine resin (60% non-volatile content)	8.07
Shellsol A	6.84
n-Butanol	3.20
Cobalt naphthenate (6% Co)	0.19
	100.00

Non-volatile content	70%
Cure schedule	30 mins at 120°C
Alkyd:melamine ratio	7.7:1

Formulations suggested by Dyno Industrier A.S.

Typical two-pack acid curing wood lacquer:

Short oil, semi-drying alkyd[1]	38.8
Isobutylated urea resin[2]	26.2
Butylated melamine resin[3]	5.8
Butyl acetate	8.0
Ethanol	6.0
Syloid 166[4]	3.5
High speed disperse, then add:	
Ethanol	11.3
Silicone oil[5]	0.2
Paint additive 56[6]	0.2
	100.0

[1] Dynotal T – 49 – EMp
[2] Dynomin UI – 16 – E
[3] Dynomin MB – 98 – E
[4] Bayer
[5] Dow

Non-volatile content	58%
Cure with 4% para toluene sulphonic acid (40% in ethanol)	
Alkyd:amino ratio	3:2
Urea:melamine ratio	3:1

Typical clear one pack pre-catalysed wood finish:

Isobutylated urea resin	19.2
Mowital B30 H	12.5
Isobutanol	50.6
Xylene	16.7
Acid butyl phosphate	1.0
	100.0

Non-volatile content	25%

Chapter VII

Polyamides

74

Chapter VII

Polyamides

INTRODUCTION

Polyamide resins used in the surface coating industry can be divided into two major categories:

> Reactive
>
> Non-reactive.

It is possible further to subdivide these classes, particularly for non-reactive resins.

The simplest representation of the formation of a polyamide is:

$$HO_2C - R - CO_2H + H_2N - R^1 - NH_2 \rightarrow HO_2C - R - CONHR^1 - NH_2 + H_2O$$

When R and R^1 are aliphatic hydrocarbons. As an example, where R is $(CH_2)_4$ and R^1 is $(CH)_2)_6$, the product is nylon 6,6.

The reaction between acid and amine groups, as a general rule, occurs fairly readily. Whilst nylons are extensible, tough durable solids, their insolubility in the majority, if not all, of the solvents commonly used in the surface coating industry (including inks) precludes their usage unless application is from a melt. Therefore, modification is required before these resins can be utilised even if some of the desirable properties of nylons have to be partially sacrificed.

Replacing an aliphatic di-acid by dimerised fatty acid (often referred to as 'dimer' or 'dimer acid') alters the properties and in particular solubility, sufficiently for usage in surface coatings and inks. Most polyamides for these industries are based on dimer acid, a few monofunctional acids and a limited selection of di- and multifunctional-amines. Because most of the raw materials used in any of the surface coating and ink polyamides are common they will be considered first.

DIMER ACID

Fatty acids can be catalytically polymerised to form a mixture of monomer dimer and trimer acid. These mixtures, which are commercially available, are normally referred to as dimer.

The fatty acids, used for dimerisation are derived from triglyceride oils, which are of natural origin, so that they vary in composition. Most fatty acids which are dimerised contain a substantial proportion of linoleic acid, the dimerisation of which can be represented in an idealised form as follows:

Conjugated linoleic acid   Dimer acid

The dimerisation of fatty acids is covered by patents to many companies, both in the USA and Europe. Catalysts used at present are types of activated clay which are filtered from the final product. Dimerisation occurs at elevated temperatures.

The composition of commercial dimer is typically:

Monomeric fatty acid	(monomer)	1−10%
Dimeric fatty acid	(dimer)	60−85%
Trimeric fatty acid	(timer)	10−20%

The amount of trimer present affects the overall functionality of the dimer and, because the trimer content is generally significantly larger than the monomer, the overall functionality is significantly greater than two. This must be allowed for when formulating polyamides, particularly if amines with a functionality greater than two are present. Even when only di-amines are present it is necessary to use monofunctional acids as chain stoppers.

Fatty acids (which are derived from oils) which are commonly used as a feedstock for dimerization, include Tall oil fatty acid (both Scandinavian and American) soya bean and castor oils.

For some specific products, di-basic acids can partially replace dimer acid. Typical di-basic acids include the following:

i) adipic acid;

ii) isophthalic acid, terephthalic acid, and phthalic anhydride.

MONOFUNCTIONAL ACIDS

Monofunctional acids are used in the manufacture of polyamides for the following reasons:

i) to reduce the overall functionality of the polymer system;

ii) to control molecular weight and hence the viscosity of the polyamide;

iii) to modify the solubility of the polyamide.

The monofunctional acid can induce alcohol or co-solvent solubility. (See later). For alcohol solubility, a low molecular weight acid would be used for chain stopping. For co-solvent solubility, a fatty acid or high molecular weight acid would be used.

The 'chain stopper' acid significantly affects the properties of the final polyamide. This is particularly true for non-reactive polyamides where the choice of chain stopper dictates whether the polyamide is alcohol or co-solvent soluble. Co-solvent polyamides are insoluble in alcoholic solvents and require the addition of aromatic or aliphatic solvents to alcoholic ones to achieve complete solubility.

The monofunctional acids which can be used as chain stoppers include:

a) Fatty acids and tall oil fatty acid in particular

b) Acetic acid

c) Isobutyric acid

d) Stearic acid

e) Octanoic and iso octanoic acids

f) Propionic acid

g) Pentanoic acid.

AMINES

For non-reactive polyamides, a di-functional amine, which is normally ethylene di-amine, is used. As a general rule only di-functional amines are used for non-reactive polyamides.

Reactive polyamides require residual amine groups for cross linking reactions, thus multifunctional amines are used either alone or mixed with di-functional ones. Both aliphatic and aromatic amines are used.

The following are some of the more commonly encountered amines.

Di-amines

a) Ethylene di-amine

b) Hexamethylene di-amine

c) Isophorone di-amine

d) Tri-methyl hexamethylene di-amine

e) Propylene di-amine.

Multi-functional Amines

a) Di-ethylene tri-amine

b) Tri-ethylene tetramine

c) Tetra-ethylene pentamine

d) Penta ethylene hexamine.

METHODS OF PREPARATION OF POLYAMIDES

Ethylene di-amine is volatile. If the temperature of the reaction mixture is raised too rapidly, excessive amine losses can occur. Thus, the general method of preparation is to prepare the amine acid salt by holding the acid amine mixture at 120—150°C under reflux for between one to four hours to 'fix' any free amine. The temperature is then raised, over ca three hours to between 200—260°C whereupon acid-amine condensation occurs with the elimination of water.

The reaction is monitored by measurement of acid and amine values and viscosity. Upon reaching reaction temperature, or increasing the temperature after fixing the amine, it is common practice to correct for amine losses or if the viscosity is too high, to add further monomeric acid.

Many of the non-reactive polyamides are formulated for equivalence of amine and acid with an allowance for amine losses. Azeotroping solvents and vacuum are frequently used to improve the rate of reaction, by assisting the removal of water of reaction. The initial process is reflux until the amine is 'fixed'. It is then converted to distillation (with a separator if azeotroping solvent is used). Steam sparging under reduced pressure is commonly used to reduce the odour of polyamides intended for use in inks.

Reactive polyamides contain an amine excess, which can be induced, by varying the process conditions, to form imidazolines. This is discussed in more detail in the reactive polyamide section.

A nitrogen sparge is used to maintain colour and assist the removal of water. An excessive flow rate will increase amine losses.

Process controls are normally amine and acid values and at a later stage, viscosity, either melt or solution. For reactive polyamides other techniques are also required (see reactive polyamide section).

Solid polyamides are generally discharged into trays, metal drums or onto the floor and after about two days they are easily broken into lumps. These lumps are then crushed or kibbled. Alternatively an endless stainless steel conveyor belt can be used for discharge to sheet the resin.

NON-REACTIVE POLYAMIDES

Non-reactive polyamides are generally solids with melting points in the range 90–150°C. They find applications in inks, hot-melt adhesives and thixotropic alkyds. The ink types of non-reactive polyamides can be further subdivided, by their solubility characteristics, into alcohol and co-solvent grades. This obviously dictates the types of inks the polyamide can be used for.

The preparation of a polyamide representing each type is outlined, as are some typical starting formulations for inks.

These are followed by a typical formulation for a polyamide for a thixotropic alkyd.

POLYAMIDES FOR INKS

Typical formulation for a non-reactive alcohol soluble polyamide resin.

RESIN A

Dimer Acid	0.5 to 0.9 equivalent
Isobutyric acid	to 1.0 acid equivalent
Ethylene di-amine	1.05 equivalent
Anti oxidant	0–2%
Silicone antifoam	0–0.2%

PROCESS

Charge the dimer acid isobutyric acid and ethylene di-amine. Heat to 140°C with agitation and nitrogen sparge under reflux. Hold for one hour until the amine salt has formed. Slowly raise the temperature to 200–260°C under distillation, collecting the water of reaction. Monitor amine and acid values. If necessary adjust for acid and amine losses. Continue processing until the desired amine and acid values are reached. These are normally both less than 6 mgKOH/g. Additional controls may include viscosity (either solution or melt) and melting point. Vacuum can be used to assist removal of water.

TYPICAL PROPERTIES OF RESIN A

Acid value mg KOH/g	6 maximum
Amine value mg KOH/g	6 maximum
Melting point °C	105–115
Solution viscosity seconds*	10–25
*(No. 2 Zahn cup at 25°C for a 30% non-volatile content IMS/ethyl acetate (3/1) solution)	

Typical formulation for a non-reactive co-solvent soluble polyamide resin.

RESIN B

Dimer acid	0.80 to 0.95 equivalent
Tall oil fatty acid	to 1 acid equivalent
Ethylene di-amine	1.05 equivalents
Anti-oxidant	0 – 2%
Silicone anti-foam	0 – 0.2%

PROCESS

A similar process to that for the non-reactive alcohol soluble polyamide would be used. Again controls would be on acid and amine values which would both be typically less than 6 mgKOH/g. Again melting point and viscosity could be used for additional controls. It is necessary to dissolve the polyamides in a mixture of butanol and toluene or xylol.

TYPICAL PROPERTIES OF RESIN B

Acid value mg KOH/g	6 maximum
Amine value mg KOH/g	6 maximum
Melting point °C	100 – 115
Solution viscosity seconds*	15 – 30 (depending upon the dimer/ TOFA ratio)
*(No. 2 Zahn cup at 25°C for a 30% non-volatile content isopropanol/SBP5(1/1) solution)	

Polyamides for Inks

The co-solvent type of polyamides require a mixture of an alcohol with an aliphatic or aromatic hydrocarbon solvent.

Occasionally there are problems with long term solution solubility. There is a tendency for the majority of polyamides to gel at low temperatures, and good anti-gelation characteristics are essential for polyamides for better quality inks. Unlike gelation in resin preparation, the gelation referred to in an ink system is when the solution solidifies or gellifies. On raising the temperature the gel disappears forming a liquid. It is not dissimilar to freezing and thawing. Gelation is more of a problem with alcohol soluble systems, because the selection of a hydrocarbon co-solvent can have a marked effect on gelation characteristics.

In addition to the temperature at which gelation occurs, the rate of recovery is important. This is normally recognised as the time taken for recovery to the fluid state at ambient temperature. Toluene tends to lower gelation temperature and improve the rate of recovery. The best compromise of properties with an alcohol is found with n-propanol, but this limits press speeds due to its low volatility (c.f. ethanol), thus if propanol is used it is only used in small amounts.

A small amount (0.5–2%) of water in the varnish improves the low temperature stability of a co-solvent based varnish.

Whilst polyamides have many beneficial properties, they can be improved by incorporating small amounts of other resins, e.g. styrene maleics, or ketone resins to improve hardness, toughness and gloss.

Nitrocellulose is the preferred additive, but the majority of co-solvent polyamides are either incompatible or have limited compatibility with nitro-cellulose. This can be improved by adjusting the solvents to include a high proportion of esters, particularly ethyl or iso-propyl acetate.

As a general rule the compatibilities of thermoplastic polyamides and different resins can be summarised as in the Versamid literature. Partially compatible resins will depend to some extent on the exact nature of the polyamide and resin concerned.

Compatible Resins include: Polymerised rosin, resinates (calcium and zinc), many maleic condensates, some phenolic, reduced phenolic and phenolic modified resins, ester gums, most terpene resins and some urea and melamine resins.

Partially Compatible Resins include: Some polyesters and alkyds (prolonged high temperature cooking is necessary) hard maleic condensates, sulphonamide resins, some acrylics, some reduced phenolics, some ketone resins, some urea and melamine resins, some hydrocarbon resins, Cellulose acetate butyrate, medium molecular weight epoxy resins, nitro cellulose.

Incompatible Resins include: Polythene, polystyrene, hard acrylics, liquid epoxy resins, ethyl cellulose, most silicones, rubber and chlorinated rubber, polyvinyl acetate, polyvinyl chloride, polyvinyl butyral and polyisobutylene.

Many suppliers of polyamides publish solubility and compatibility data.

Packaging Inks

Many polyamide resins find application in packaging inks either gravure or flexographic.

Obviously for flexographic inks the nature of the solvent or solvent mix is critical to minimise or obviate the swelling of rubber stereos. This may limit the choice of polyamide to an alcohol soluble or reducible one.

Many of the requirements of a good flexographic packaging ink illustrate the reasons why polyamides have found such a large use in these applications.

The adhesion of polyamides is unsurpassed and this, combined with their flexibility, makes them ideal for incorporation into flexible packaging inks and indeed for some substrates like inadequately treated polythene film only polyamide based resins will perform satisfactorily. Adhesion on paper, metal foils and cellulosic films is also excellent.

Combined with the adhesion of polyamides is outstanding resistance to the majority of the chemicals to which a packaging ink is normally exposed. This includes water, detergents, grease, oil etc. Water resistance is improved by the addition of nitrocellulose.

Inks dry (harden) by solvent evaporation, unlike alkyd based inks which react through oxidation of the unsaturated bonds or other reactive binder systems. Therefore not only must the ink be soluble in the solvent system it is applied from, but it must also release the solvent rapidly to allow high printing speeds. Failure to release means that solvent is retained in the ink film when it is wound on to the reel after printing. The solvent is trapped and will cause odour, contamination and poor print performance, including blocking.

Odour is very important, particularly for packaging inks for confectionery and food applications. The choice of pigment and solvent can affect odour. However, polyamide manufacturers are all striving continuously to reduce the odours of polyamide based resins.

Gravure methods apply approximately twice the film weight which flexographic ones do, thus solvent release and anti-blocking characteristics have to be even better than those for flexographic inks. However, the facility to use co-solvent inks for gravure printing, allows the resin formulator to build these properties into the resin. There can be additional improvements in resistance properties, particularly water resistance, as a result of these changes.

Inks based on Co-solvent Polyamides

Co-solvent grades of polyamide can be used in flexographic systems where the stereos are more resistant to solvent than in some other presses, or where damage to the stereo is considered less important than obtaining the maximum properties of the polyamide resin.

Two starting formulations for flexographic inks based upon Co-solvent Polyamide Resin B are:

WHITE INK

Resin B	20
Titanium di-oxide	33
Supropal AP (Ex-BASF)	5
Isopropanol	20
n-Propanol	6
SBP 3	14
Polyethylene wax	2
	100.0

COLOURED INK

Resin B	20
Organic pigment	12
SK 100 (Ex-Huls)	5
Isopropanol	31
n-Propanol	10
SBP 3	20
Polyethylene wax	2
	100.0

Care must be exercised in the choice of the organic pigment because polyamides have residual amine groups which can react with groups in the pigment. Manganese lakes are sensitive to amines, and PMTA pigments tend to diffuse through polyethylene and polypropylene films if only polyamide is present. This can be reduced by the incorporation of some nitrocellulose.

Inks based on Alcohol Soluble Polyamides

Some starting formulations for inks based upon alcohol soluble polyamides are as follows, using the alcohol soluble Resin A as an example:

WHITE INK

Resin A	20
Titanium di-oxide	33
Nitrocellulose	5
74 OPMS (99% IMS)	30
Isopropanol	10
Polyethylene wax	2
	100.0

COLOURED INK

Resin A	20
Organic pigment	12
Nitrocellulose	5
74 OPMS (99% IMS)	40
Isopropanol	21
Polyethylene wax	2
	100.0

Modifying resins include; resin esters for improving adhesion and hardness, maleic and ketonic resins for upgrading gloss and styreneated maleates for improving the toughness of the film. However, the best properties are obtained when polyamides are combined with nitrocellulose. Normally a low viscosity alcohol soluble grade of nitrocellulose would be used, often in combination with an ester solvent, which will give a beneficial lowering of the viscosity of the varnish.

Frequently small differences between polyamides, particularly water resistance and gloss, are masked by the contributions to the overall properties from the nitrocellulose component.

Polyamide and nitrocellulose can be chipped (i.e. two roll milled) with pigment. This gives good dispersion of pigment and normally excellent gloss. The ratio of nitrocellulose to polyamide will also, of course, effect gloss, and other properties.

The commonly used range of ratios of polyamide/nitrocellulose lies between 1:2 and 4:1. Alternative methods of incorporating nitrocellulose are milling or high speed mixing.

NITROCELLULOSE MODIFIED INK

Resin A	15
Organic pigment	12
Nitrocellulose	10
74 OPMS (99% IMS)	30
Isopropanol	20
Ethyl acetate	11
Polyethylene wax	2
	100.0

Inks containing polyamide, nitrocellulose and metallic pigments are unstable due to the decomposition of nitrocellulose. Oxides of nitrogen are evolved, and gelation of the ink occurs.

Varnish stabilities in different solutions and resistance to gelation at low temperatures are important. Some suppliers of polyamides publish individual compatibility and varnish stability data.

The same care in pigment selection as for co-solvent grades of polyamides is required for inks containing alcohol soluble grades.

Varnishes based on Polyamides

Many polyamides find application in varnishes or lacquers which are, in essence, solutions of polyamide resin, perhaps with modifying resins, and a few additives like wax, to improve slip and scuff resistance.

The majority of the varnishes are applied over the print to protect the ink film. They may also serve other functions (e.g. cold or heat seal coatings). The first varnish outlined is normally called an overprint varnish and properties required include high gloss, good water and everyday chemical resistance, hardness, toughness, good adhesion and flexibility.

Low colour is important for 'clear' pastel shade and white inks. As thick a film as possible is required, consistent with colour, application method, film defects caused by too high a film weight and solvent release. Obviously gravure printing is the preferred technique.

For heat seal coatings the thermoplastic nature of polyamides coupled with their excellent adhesion makes them suitable, but their relatively low melting points is a major drawback where temperatures higher than about 140°C are employed. Blocking is a major problem.

For cold seal coatings, which are gaining importance in packaging thermally sensitive foodstuff, like ice lollies, or chocolate, the polyamide varnish has to stop the print from becoming damaged by adhesive when the packaging is rolled. The coating must also act as a release coating, but migration must not occur from the polyamide to the latex based adhesive. Frequently wax is added to coatings intended for cold seal coatings.

Solvent release has an important effect on film performance. Retained solvent will affect the properties of the film. With higher technology applications and higher web printing speeds, the demands on good solvent release are increasing all the time. Odour is also increased with decreasing solvent release.

FLEXOGRAPHIC OVERPRINT VARNISH

Wolfamid 108	35
Isopropanol	35
SBP 3	30
	100.0

GRAVURE OVERPRINT VARNISH

Wolfamid 108	35
Isopropanol	40
Toluene	25
	100.0

POLYAMIDES FOR THIXOTROPIC ALKYDS

Typical formulation for a polyamide for thixotropic alkyds:

Dimer acid	0.80 – 0.95 equivalent
Tall oil fatty acid	to 1 acid equivalent
Ethylene di-amine	1.05 equivalent
Silicone anti-foam	0 – 0.2%

The process and formulation are similar to that for the non-reactive co-solvent soluble polyamide resin.

PROCESS

Polyamides can be reacted with alkyd resins to give a thixotropic (i.e. non drip) consistency. Polyamides can be formulated to give various gel strengths (i.e. different degrees of non-drip characteristics). Gel strength is influenced by many factors which include the amount of polyamide, the temperature and time used for reacting ('cooking') the polyamide with the alkyd, the nature of the alkyd and the properties of the base alkyd.

The majority of polyamides for thixotropic usage are designed to be incorporated into medium or long oil length alkyds. The market for thixotropic short oil alkyds is minimal because short oil alkyd based paints are generally spray applied. The incorporation of a special thixotropic grade polyamide resin into a short oil alkyd will reduce or eliminate sagging.

For the optimum performance of a 60% oil length alkyd based upon soya bean oil, pentaerythritol and phthalic anhydride the following properties of the alkyd resin are required: Acid value — 8–12 mg KOH/g; Hydroxyl value — 40–60 mg KOH/g; Viscosity — 6–9 poises (55% non-volatile content in white spirit) at 25°C.

For a higher viscosity, maleic anhydride or bodied oil could be used.

Aromatic solvents in the base alkyd can reduce the thixotropic (gel) strength. Any Xylene (or other aromatic solvent) used as an azeotroping solvent must be removed before the polyamide is added. Less than 0.5% aromatic solvent must remain in the alkyd, to avoid loss of gel strength.

Normally 4−9% by weight of polyamide on the base alkyd would be added to the base alkyd at temperatures between 190−230°C. It is believed that an ester/amide interchange occurs whereupon polyamide fragments are incorporated into the overall alkyd structure. As the reaction proceeds, the solubility of the reaction mixture, in either ester solvents or white spirit, increases with increasing degradation and incorporation of the insoluble polyamide resin. The undissolved resin causes haziness and the degree of clarity as determined by a Nephelometer can be used for monitoring the reaction. At any temperature, the degree of thixotropy increases to a maximum, and then starts to decrease.

The rate of increase and decrease in thixotropy is determined by the temperature and duration of the reaction. Different shaped, gel strength and time curves are obtained for each temperature.

It is normally necessary to obtain a maximum level of clarity at the maximum level of thixotropy, and this is normally achieved by reacting at the lowest temperature practical.

Upon achieving the desired gel strength and clarity the reaction mixture is cooled, as rapidly as possible, to 130−150°C to quench any further reaction.

It is difficult to assess the effect on thixotropy of a polyamide on the laboratory scale. It is necessary to 'fine tune' alkyd formulations in the plant. Using the same polyamide and alkyd formulation it is feasible for one company to obtain a satisfactory thixotropic alkyd and another company (or even different plant in the same company) be unable to manufacture a thixotropic alkyd satisfactorily.

REACTIVE POLYAMIDES — EPOXY RESIN CURING AGENT

Reactive polyamides are generally viscous or highly viscous liquids, with fairly high residual amine values.

Reactive polyamides generally cover a range of polyamide, polyamino amine and cyclo aliphatic amine curing agents, all capable of undergoing cross linking reactions with compounds containing epoxy groups, to form tough resistant cross linked films, giving outstanding protection and adhesion. These are normally two pack systems which means that epoxy and polyamide resins are mixed prior to application.

This chemical reaction occurs even at room temperature. The time between mixing the two components and the mix becoming unusable (due to viscosity increase) is commonly referred to as 'pot life' at that temperature (normally ambient). The lower the temperature, the slower the reaction and under normal conditions the reaction is not instantaneous. Many reactive polyamides have viscosities which are too high for ease of handling or method of application, thus many commercially available reactive polyamides contain solvent. Normally xylene or toluene are used. A typical level would be 20−40% of xylene and the resin would end in the letter X (e.g. Versamid 115X, Wolfkur 1002X, Casamid 172X).

The amines normally used for reactive polyamides are multi-functional. This enables them to retain amine functionality, after some of their groups have formed amides with carboxylic acid groups. This obviously forms a polyamide with a multitude of potential cross linking sites.

Reactive Polyamide Preparation

Typical formulation for a reactive liquid polyamide resin:

Dimer acid	0.7 – 0.95 acid equivalents
TOFA	to 1 acid equivalent
Tri-ethylene tetramine (TETA) – (or other multi-functional amine)	2 – 3 amine equivalents

PROCESS

Charge dimer and TOFA and warm to 50°C. Charge TETA and carefully raise the temperature to 180 – 220°C. Control on amine value and viscosity. Higher temperatures can be used if required.

Upon completion of the reaction, xylene or other solvent can be added. A typical level would be 70% non-volatile content.

Multifunctional amines which can be used include di-ethylene tri-amine, tri-ethylene tetramine, tetraethylene pentamine and penta ethylene hexamine or mixtures of any or all of them. Cost, functionality, viscosity and film performance (i.e. cross link density) are of course affected by altering the amines.

Imidazoline formation is encouraged by alteration of processing conditions. Imidazolines affect the rate of cure, with faster cure rates being achieved with higher levels. There are differing opinions as to the importance of imidazoline content and to the change in content between manufacture and application. Imidazoline groups are characteristic of many reactive polyamides and many users insist on certain imidazoline levels.

The structure of epoxy resins has been reviewed in Chapter V. The remainder of this chapter is concerned with the epoxy amine reaction and formulations for 'two pack' systems.

There are four traditional reactive polyamide resins which are readily available from the majority of polyamide suppliers. These resins are variations of the above formulation. Ratios of dimer to TOFA are varied as are the nature, the level and the mixture of different polyfunctional amines. Typical properties of these four polyamide resins are as follows:

Property	Resin A	Resin B	Resin C	Resin D
Amine value (mg KOH/g)	80 – 100	200 – 240	290 – 320	360 – 400
Viscosity (poises @ 40°C)	*too viscous	450 – 800	74 – 150	25 – 65

*Viscosity is normally measured in a 50% non-volatile
content solution in butanol/toluene (1/1) at 25°C.

In addition to the reactive polyamides above, there is a wide range of low viscosity polyamino-amide resins formed by the reaction of fatty acids and polyfunctional amines rather than predominantly dimer. Polyamino-amide resins have a wide range of properties. Typically amine values vary from 250–450 mg KOH/g and viscosities from 1–50 (PRS @ 25°C). Imidazoline contents also vary.

THE EPOXY AMINE REACTION

Both primary and secondary amines react with epoxy groups. Primary aliphatic amines react at room temperature with the epoxide group, forming secondary amines which are capable of further reaction with another epoxy group, to form a tertiary amine link between two epoxy resin molecules. Representing the non-reactive part of the epoxy resin as: —▭—

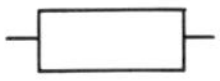

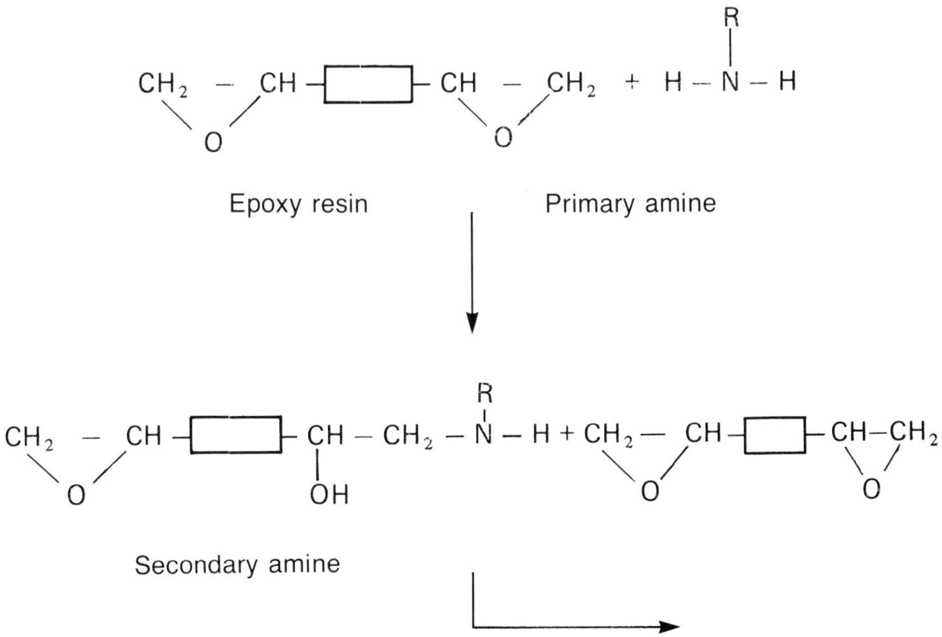

Epoxy resin   Primary amine

Secondary amine

$$CH_2 - CH \boxed{} CH - CH_2 - \underset{\underset{\displaystyle OH}{|}}{\overset{\overset{\displaystyle R}{|}}{N}} - CH_2 - CH \boxed{} CH - CH_2$$

Tertiary amine

The reactions of the hydroxyl groups formed, and those initially present in the epoxy resin with epoxide groups are suppressed, when secondary and primary amines are present which react preferentially.

The tertiary amines formed by these reactions are too sterically hindered to act as a catalyst for the hydroxyl-epoxy reaction. However, many of the tertiary amines present in very small quantities in commercialy used amines, can act as catalysts for the epoxy hydroxyl reaction, and this reaction is one of the reasons why, in practice, a slight epoxy excess over amine is preferred.

This ensures, in theory at least, that all the amine groups react, and any residual epoxy groups have the facility for reaction with hydroxyl groups.

The use of di-amines, or higher functionality amines, enables extensive three dimensional cross linked networks to be formed, giving a tough, resistant film.

In practice, primary aliphatic di-amines are too volatile to be satisfactorily used as curing agents. Higher molecular weight polyamines are therefore used. Many of the commoner types are reaction products of dimerised fatty acid and multi-functional amines, containing residual amine groups without inherent volatility, and are used in conventional 'two pack' systems.

Aromatic amines are not as reactive as aliphatic ones and require high temperatures (ca. 200°C) before appreciable cure occurs. One pack systems with long shelf lives can be formulated from aromatic amines at room temperature.

One of the drawbacks of the conventional 'two pack' reactive polyamides is high viscosity. For this reason ranges of high molecular weight, low viscosity, aliphatic polyamino amine curing agents are available. Polyamide curing systems are of lower toxicity, slower cure rate, lower exotherm and longer pot life compared to polyamine systems. The dimerised fatty acid based curing agents give films of excellent flexibility, but they tend to be softer than polyamine cured ones. However, the latter require more exact formulation for optimum cure, unlike the reactive polyamides, which are exceptionally tolerant to formulation and will cure satisfactorily over a wide range of mixing ratios with epoxy resins. For applications in two part adhesives, film hardness is not normally of paramount importance. Adhesives are considered separately, but the reactions outlined here are applicable to both.

Another reaction which can occur to give cross linked networks is etherification through the epoxy hydroxyl reaction. This occurs through an alkoxide ion formation which is induced by a tertiary amine catalyst (e.g. R_1, R_2, R_3N) and these are often present in small quantities as impurities. The alkoxide ion is then capable of providing a site for chain branching and subsequent cross-linking, forming an extensive three dimensional structure. Secondary hydroxyl groups may be present in the epoxy resin and are also generated as a result of the epoxy amine reaction. The epoxy hydroxyl reaction can be summarised as follows:

$$R - \underset{\underset{H}{|}}{\overset{\overset{OH}{|}}{C}} - R' + CH_2 - CH - R'' \longrightarrow R - \underset{\underset{H}{|}}{\overset{\overset{R'}{|}}{C}} - O - \underset{\underset{H}{|}}{\overset{\overset{H}{|}}{C}} - \underset{\underset{OH}{|}}{\overset{\overset{H}{|}}{C}} - R''$$

THE USE OF POLYAMIDES
IN CURING EPOXY RESINS

There is a multitude of materials which will react with epoxy resins, and, depending upon the number of functional groups per molecule, cross linked networks can be formed eventually.

One of the commonest types of chemical reaction used is that between amine (either primary or secondary) and epoxy groups, and this method of curing (which has already been outlined) is the long established approach used in 'two-pack' paint systems.

The amine curing systems available to the formulator can be divided into three general categories:

  i) Low molecular weight amines, which are simple amine molecules, generally termed free amines.

 ii) Adducts, however formed, represent intermediates in the curing process. These are generally epoxy resins, reacted with an excess of amine containing compounds, leaving residual amine groups for further reaction.

iii) Polyamide resins containing an excess of amine groups which are capable of undergoing further reaction with epoxy groups. These are classified as reactive polyamides, which can include traditional reactive polyamide resins, based upon dimerised fatty acids and lower viscosity alternative polyamino amine and cyclo aliphatic amine curing agents.

Each category of curing agent has its own advantages and disadvantages. Free amine curing agents tend to be volatile with irritating vapours, the inhalation of which could be harmful. Because of the low molecular weight of the amine, small accurate quantities of curing agent are usually required and this may present a problem at the application stage. This is particularly so since a small mischarge may significantly affect the properties of the final film. The cure rate is very fast, but the pot life is generally short. Provided the films are aged before exposure good chemical resistance can be obtained, but there is a tendency for the films to bloom.

As expected, the properties of films cured by adduct curing agents are generally between those of free amines and polyamides.

Reactive polyamides are generally low volatility, relatively low odour, viscous liquids, with the amine functionality attached to the polyamide molecules, thereby making them much less harmful than free amines.

The flexibility of polyamide cured films is superior to free amine or adduct cured systems. Curing times tend to be longer, but this has the advantage of extended pot life. For optimum water resistance, polyamides would be used in preference to other curing agents. These systems can be applied under damp conditions, unlike free amines.

Optimum solvent resistance is obtained with either amine or adduct cured systems, but polyamide systems generally give satisfactory resistance.

Due to the molecular weight of the polyamides and the number of functional (reactive) groups per molecule, similar weights of polyamide and epoxy resins are used (usually 30–120 pts polyamide per 100 pts epoxy resin). This has the additional advantage of formulation tolerance to mischarging the mixing ratios and yet obtaining a cured film with adequate performance properties.

Reactive polyamides are used in protective coatings for structural steelwork and corrosion resistant coatings. Unlike other curing agents, reactive polyamides are inherently corrosion resistant and are preferred for such applications.

Polyamide resins are generally more tolerant of poor preparation of substrate, and form an impervious barrier over it.

Polyamide resins do not require ageing, and are much less likely to bloom than free amine cured systems.

HOW MUCH POLYAMIDE RESIN TO USE?

The selection of the polyamide and epoxy resin determine the amounts to use in starting point formulations.

One of the easiest ways to compare the number of amine and epoxy groups present in any formulation is to relate them to amine and epoxy values, both expressed as acid equivalents in mg KOH/g.

The epoxy value is obtained from the E.E.W. (or E.P.W. or E.M.M.) as follows:

$$\text{Epoxy value} \quad = \quad \frac{56100}{\text{E.E.W.}} \quad \text{mg KOH/g}$$

Traditionally the resin industry uses two methods of determining amine value and the values obtained depend upon whether hydrochloric or perchloric acids are used. The hydrochloric acid method which is commonly used and upon which most of the values quoted in the literature are based, titrates all the primary and some (but an unknown amount) of the secondary amine groups. The perchloric acid method titrates all the primary and secondary amines. Generally speaking, most formulators tend to use a slight excess of epoxy groups. The reasons include epoxy hydroxyl reaction and the fact that the amine value used for the calculation is frequently lower than the true one relating to the number of reactable groups. Thus, the real ratio is often nearer a theoretical one than would appear at first sight.

Obviously, the formulator must adjust the formulation to optimise the properties and performance required.

The following example illustrates the calculation of theoretical weights of epoxy and polyamide resin required:

Epoxy resin E.E.W. of say 190 gives an epoxy value of 295 mg KOH/g. Polyamide resin A of amine value of say 220 mg KOH/g. Then for every 100 g of epoxy resin:

$$\frac{100}{220} \times 295\,\text{g of Polyamide resin B} = 134\,\text{g}$$

are needed for complete reaction which represents a ratio of polyamide/epoxy of $4:3$. Thus, the calculation could be written:

$$\text{weight of polyamide resin} = \frac{\text{weight of Epoxy resin} \times \text{Epoxy value}}{\text{Amine value}}$$

$$= \frac{56100}{\text{E.E.W.}} \times \frac{\text{weight of Epoxy resin}}{\text{Amine value}}$$

When solutions are used it must be remembered that amine values quoted refer to that of the solution and not the non-volatile part. The figures are given this way to enable a weight of solution to be calculated rather than having to correct for non-volatile content.

The effect of polyamide/epoxy ratio on final film properties can be summarised by considering the properties required.

Hardness and degree of cure increase with increasing epoxy content of the coating as does aromatic solvent resistance. A better coloured film is generally obtained because of the low initial colour of the epoxy resins.

Increasing the polyamide content increases flexibility and resistance to alcoholic solvents but the film becomes softer. High polyamide content systems tend to be used for adhesives rather than coatings.

However, it must be stressed that the paint formulator must determine experimentally the optimum epoxy/polyamide ratio required for any particular formulation and application. This introduction and the formulations given are only to be used as guidelines.

POT LIFE

Pot life depends upon many factors which also similarly affect rate of cure and include:
   a) Temperature;
   b) Epoxy/polyamide ratio;
   c) Total mass of reactants mixed;
   d) Type of polyamide;
   e) Nature and level of solvent (if any);
   f) Additives.

The effect of each factor on the pot life will be briefly considered. The pot life is timed from when the initial reactants (epoxy/polyamide) are intimately mixed to give a total of 250 grammes. The pot life is then taken as the time required for the viscosity to double.

Temperature

Increasing temperature decreases pot life as would be anticipated from the fact that increasing temperature increases rate of cure. For some applications, mixtures are stored in cold rooms/refrigerators until required to give a pot life to last a working shift.

Epoxy/Polyamide Ratio

Increasing the epoxy content for the same polyamide may alter the pot life. The ratio change will also affect viscosity of the mix which might mask the magnitude of the effect.

Total Mass of Reactants Mixed

The epoxy/polyamide reaction is exothermic, therefore the pot life will depend upon the mass of material present. The larger the mass the shorter the pot life.

Nature of Polyamide

Obviously, the chemical nature of the polyamide will affect pot life particularly as multifunctional amines are used in their preparation. This gives rise to products with different degrees of branching and the higher the branching the quicker a cross linked film could be formed and the potentially shorter the pot life. Comparisons are difficult because different polyamides use different levels of epoxy resin. The use of polyamino amines instead of dimeric fatty acid based polyamides will also significantly alter cure rate.

Nature and Level of Solvent

Solvent type and level and in particular the former, will have significant effects on pot life rate and degree of cure. This is considered later. Alcohols tend to reduce pot life, whereas ketones and esters can extend it. The concentrations of the reactive components are affected by level of solvent and the more dilute the system the slower the rate of reaction. More noticeable is the affect of viscosity on pot life, but this and concentration are inter-related.

Additives

Sometimes additives are used to alter cure rate, and these will obviously affect pot life.

DEGREE AND RATE OF CURE
AND FINAL FILM PROPERTIES

Degree of cure can be determined by an M.E.K. (Methyl Ethyl Ketone) solvent extraction test. This involves exposing the test film and blanks to refluxing M.E.K. and determining:

  i) The amount of resin in the M.E.K.: and

  ii) The percentage weight loss of the film after evaporation of solvent.

A minimum exposure of the film to solvent of one hour is recommended. The rate of cure is determined from the degree of cure at varying time intervals. Many of the factors affecting pot life also affect degree and rate of cure.

If the degree of cure after 48 hours (by which time essentially full cure has been attained) is plotted against level of resin B in resin B/Epoxy resin (E.W. 500) mixtures, a curve similar to that below is obtained:

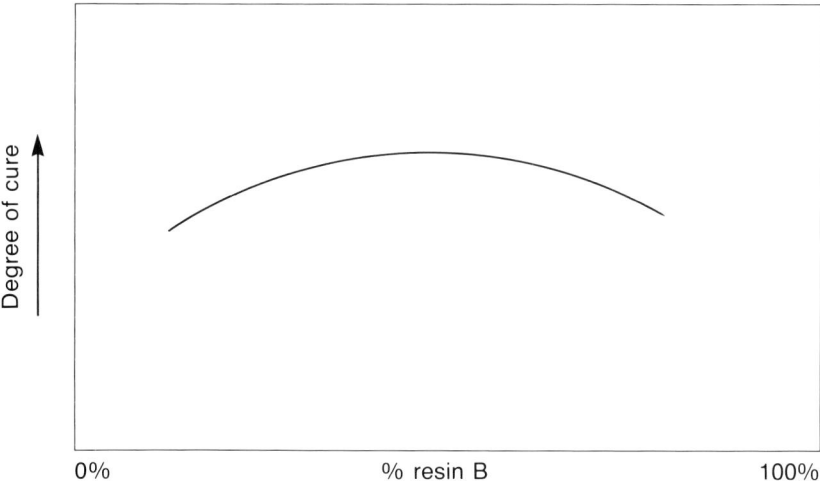

This illustrates that the same degree of cure can be obtained for a wide variation in polyamide resin content and shows the tolerance of the polyamide curing systems to mixing ratio.

Comparison of the series of polyamide resins A−D show that resin A is least tolerant (with respect to degree of cure) to variations in polyamide content, and resin D is most tolerant with minimal differences being found over a wide composition range. These comparisons and those of rate and degree of cure, are summarised in the following table:

COMPARISON OF RESINS A to D

	Theoretical amount of Polyamide*	Tolerance mixing ratios	Rate of Cure	Degree of Cure
Resin A	i	i	i	i
	n	n	n	n
Resin B	c	c	c	c
	r	r	r	r
Resin C	e	e	e	e
	a	a	a	a
Resin D	s	s	s	s
	e	e	e	e
	s	s	s	s

For a given amount of epoxy resin of similar E.E.W.

   Increasing the amount of polyamide (from a very low level) will increase the cure rate which then reaches an optimum value over a range of resin contents, but depending on the actual polyamide used, there is a wide variation of ratios over which constant cure rate can be achieved. Further increase in polyamide content will decrease the degree of cure.

   Temperature is one of the most significant factors affecting rate of cure. Films cured at 15°C and 25°C differ significantly in the amount of M.E.K. extractables over a 24 hour period from application. The final films will differ slightly in degree of cure. Obviously this means a better performance of air cured coatings can be obtained when the ambient temperatures are higher. The difference of temperature affecting degree of cure can be further demonstrated by comparing tensile strengths (which would be appropriate for adhesive application against temperature of cure).

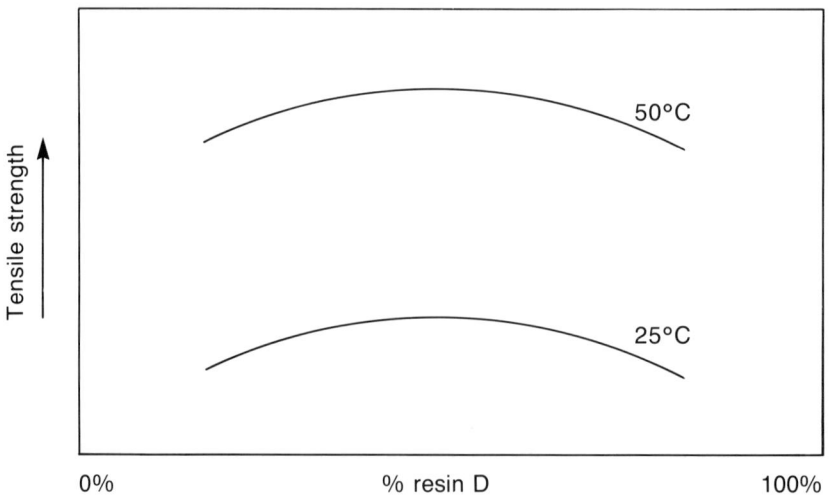

The effect of temperature is again illustrated by the stoving schedules recommended by Shell for Epikote 1001 based formulations:

FORCE DRYING		
38°C	:	1½ – 2½ hours
43°C	:	1 – 1½ hours
50°C	:	¾ hour
60°C	:	½ hour

STOVING		
93°C	:	¾ hour
120°C	:	½ hour
150°C	:	¼ hour
177°C	:	$\frac{1}{6}$ hour

Force dried films continued curing under ambient conditions over the following 24−48 hours giving films with optimum properties.

The types and level of solvent are both important factors affecting cure rate and properties of cured film. The level of solvent will affect dried film thickness, initial concentration of reactants and solvent release properties.

The higher the molecular weight of the applied film on application the higher the possibilities of rapid cure, particularly if a touch (or sand) dry test is used. This is frequently called a tack-free time.

Accelerators can be used to increase rate of cure and these include tri-phenylphosphite (T.P.P.), phenol, resorcinol and modified phenol. Discolouration will occur on curing with some of these products but for many protective, corrosion resistant coatings this is of little or no consequence. T.P.P. also additionally reduces the application viscosities of solventless coatings.

PIGMENT SELECTION

Pigment selection must be made, remembering that the polyamides resins have high amine values (generally 200−400 mgKOH/g) and any pigment which may react or become unstable in these conditions must be avoided. Where the pigment which has to be used is suspect, then it must be dispersed in the epoxy resin, thereby minimising its exposure to any free amine groups. Pigment suppliers are able to advise on the suitability of any particular pigment.

Furthermore, because epoxy polyamide systems are generally employed where exceptional durability and resistance are required, it seems pointless to use pigments which may fade or be attacked by the environment. The pigments should not be the weak spots of the film. Therefore, light fast, chemically resistant pigments are recommended. Epoxy, polyamide systems have a tendency to chalk and low chalking grades of pigment are recommended.

Conventionally, pigments tend to be dispersed in the epoxy component. This safeguards against any possible side reactions. However, with their high amine values, polyamide resins have good pigment wetting characteristics.

Zinc based primers are generally prepared by dispersing zinc dust in the epoxy component. However, care must be exercised with the zinc dust and it is recommended that raw materials are checked before incorporation into production batches of paint to avoid any film defects or subsequent hydrogen evolution due to the zinc water reaction. It is essential that any materials coming into contact with zinc are dry.

SELECTION OF EPOXY RESIN AND CURING AGENT

The choice of epoxy resin and, indeed, polyamide resin, partially depends upon whether a solventless coating is required. Solventless coatings restrict the formulator to liquid low molecular weight epoxy resins with either polyamides or polyamino amine curing agents. Cyclo aliphatic amines could be used. Care must be exercised in using some liquid grades of epoxy resin because of the possible increased hazards associated with some of the reactive diluents like butyl glycidyl ether, which are present in some liquid epoxy resins to lower the viscosity. Viscosity decreases with increasing temperature but pot life decreases rapidly. Therefore, there is a practical limit to the temperature the reaction mixture can be heated to before application.

Solvent coatings can utilise the harder higher molecular weight epoxy resins in addition to the liquid grades. High viscosity reactive polyamide resins diluted in various solvents can also be used in addition to those listed above.

CHOICE OF SOLVENT

The nature of the solvent or as is more usual solvent mixture will affect degree and rate of cure, as well as final film properties (depending upon the end application). The method of application often dictates the solvents which can be used. For spray systems, mixtures of high volatility solvents are used and these would typically include isopropanol, butanol, toluol, M.E.K. and M.I.B.K. and xylol. If the flash off rate is too fast to allow film levelling, then a small proportion of a lower volatility solvent could be added.

A typical solvent mix for a spraying paint is:

Epoxy Component	% wt
Methyl ethyl ketone	25
Methyl isobutyl ketone	25
Toluene	100
Polyamide Component	% wt
n-Butanol	40
Toluene	40
Methyl oxitol	20

For brush systems, it is important that the can viscosity is only minimally affected by solvent evaporation and generally solvents of much lower volatility are used. Typically, these would include Aromasols, Shellsols, Carbitols and Cyclo Hexanones.

A typical solvent mix for a brushing paint is:

Epoxy Component	% wt
Methyl oxitol	80
Shellsol A	20
Polyamide Component	% wt
n-Butanol	40
Xylene	40
Methyl oxitol	20

As a general rule, polyamides dissolve in less polar solvents than epoxy resins and care must be taken to ensure that the solvent mix present in the mixed compound is sufficiently strong to solvate the epoxy resin, otherwise some degree of precipitation and inhomogeneity will occur. Additionally, the molecular weight of the reaction mixture increases whilst in the pot before application and the solvent must be capable of solvating this material as well. Thus, a stable solution is required throughout (and in excess of) pot life and during the evaporation from the film to avoid differential precipitation of resin and subsequent film defects and inadequate performance.

The method of solvent removal from the film also dictates the type of solvents which can be used. To obtain maximum performance from the coating, it is necessary to remove the solvent. Thus, air dried films should contain volatile solvents or be used in applications where high performance is not required. If water soluble solvents are used then stoving is required or, for air drying systems, solvents of high volatility are required, otherwise entrained solvent will lower water resistance.

It is also important to dissolve certain resins in different solvents. Reactive polyamides can form complexes over typical storage periods with ketones or esters and these will reduce cure rate. Therefore, only the epoxy component should be dissolved in these solvents.

For some solvent systems, better film properties are obtained if the mix is allowed to stand for approximately 20−30 minutes prior to application to allow complete solution and stability in the new solvent system. This would also allow initial chemical reaction to commence, which will result in slightly higher molecular weight materials being applied and subsequently better film properties.

Di-acetone alcohol should be avoided at all costs because amine containing resins degrade it forming acetone.

REACTIVE DILUENTS

An alternative to using solvent to reduce the viscosity of the coating is to use reactive diluent. This is a compromise. Viscosity is reduced but generally not by as much as by the addition of an equivalent amount of solvent. The diluent enters into film formation but the properties of the cured films are generally inferior to those without diluent.

Diluents include glycidyl ethers. However, they should be used cautiously because of potential toxicity hazards associated with some of them. Some grades of epoxy resin contain a proportion of reactive diluent as supplied by the manufacturer.

An alternative approach is to use low viscosity polyamino amine curing agents as diluents, and these participate in both film formation and cross-linking reactions, both being curing agents in their own right.

APPLICATIONS

The mixing of components must be conducted in such a manner as to minimise any aeration. After mixing, the mixture is generally allowed to stand. Ideally, and particularly for viscous systems where high performance is required, degassing may be necessary. Application of a vacuum or careful film application is generally sufficient, otherwise air bubbles may possibly be entrained in the coating (which is increasing in viscosity and reducing the possibilities of their release) which would give rise to 'weak points' in the final film.

It is preferable that surfaces of substrates are clean, free from grease, etc., and loose material before application of the coating. For optimum performance of metal coatings, pre-treatment of the metals are recommended. Iron and steel are normally phosphated, whereas aluminium and copper are chromated. Sand or shot blasting is desirable for structural steelwork to obtain maximum performance of the coating. However, some of the major advantages of the epoxy polyamide coating systems include:

1. Tolerance to residual water on the substrate, enabling application under damp conditions. Polyamino-amides can be used where there is a high level of moisture present. Many applications requiring high tolerance to water during curing also require corrosion resistant, high performance coatings (e.g. oil rigs). Polyamide/epoxy resin systems can be applied as an aqueous emulsion to lower viscosity.

2. Tolerance to poor surface preparation. Unless there are extremes of contamination, epoxy polyamide systems will cure to some extent to give a serviceable film, even though it need not be the best theoretically obtainable.

3. The ability to coat rust and act as a moisture penetration barrier.

4. Excellent adhesion to a wide range of substrates.

5. The epoxy polyamide coating is, in itself, a corrosion inhibitor and this reduces/stops further rusting, when used for re-coating. It also upgrades the performance of corrosion inhibiting pigments and frequently allows inexpensive pigments to be utilised.

6. Durable coatings, particularly in corrosive atmospheres. However, chalking is a problem and results in loss of gloss. Epoxy polyamide systems are better than epoxy amine ones for gloss retention. Many of the polyamide epoxy systems are used in primer applications where chalking is of less importance and adhesion and corrosion resistance are of paramount importance.

Major areas of usage of reactive polyamides include 'two pack' adhesives, sealants and paints. Paints include tank linings, corrosion resistant primers and top coats including metallic (zinc or lead) systems.

These products also find applications in the electronics industry as encapsulents (i.e. potting).

FORMULATIONS FOR EPOXY/POLYAMIDE COATINGS

RED LEAD PRIMER

Component A	Parts by weight
Non-setting red lead	73.33
Asbestine	6.67
Resin B (70% in xylene)	11.33
Xylene	8.00
Cyclo hexanol	0.67
	100.00

Component B	
Epoxy resin (E.E.W. 500)	52.0
Xylene	28.0
n-Butanol	20.0
	100.0

Mix component A with component B in the ratio 3:1 (A:B). The mixture has a pot life of 22 hours at ambient temperature. This primer is recommended for high duty corrosion protection of structural steelwork and chemical plants.

A 52 micron film (dry thickness) was applied to a mild steel panel and allowed to cure for 24 hours at room temperature. The panel was cross hatched and exposed to salt spray testing (BS 3900) for 1200 hours. There were no signs of rusting.

ZINC RICH PRIMER
(Formulation 1)

Component A	Parts by weight
Epoxy resin (E.E.W. 500)	34.7
Anti-settle additive	3.0
Cellosolve	20.0
Methyl isobutyl ketone	20.0
Xylene	22.3
	100.0

Component B	
Resin B	37.9
Cellosolve	20.0
Methyl isobutyl ketone	20.0
Xylene	22.1
	100.0

Component C	
Zinc dust	100.0

Mix component A and B in the ratio 10:4. The add zinc dust (component C) to give 80% zinc in the final mixture. This will give 92% zinc in the cured film.

ZINC RICH PRIMER
(Formulation 2)

Component A	Parts by weight
Epoxy resin (E.E.W. 500)	42.0
Xylene	30.0
Butyl Cellosolve	18.0
Methyl isobutyl ketone	10.0
	100.0

Component B	
Resin A	42.0
Xylene	40.0
Butyl Cellosolve	18.0
	100.0

Components A and B should be blended together (1:1) and allowed to stand for an induction period of ½−1 hour. Zinc dust should then be blended in at a ratio of 850 parts zinc dust to 200 parts of the epoxy polyamide blend. The formulation will give 91% by weight of zinc in the dry film. This coating will give excellent long term protection to steel after top coating.

ZINC RICH PRIMER
(Formulation 3)

Component A	Parts by weight
Part 1	
Zinc dust	79.7
Bentone 27	0.8
Butyl Cellosolve	2.4
	82.9
Part 2	
Epoxy resin (E.E.W. 500)	3.2
Xylene	11.3
Butyl Cellosolve	1.5
Methyl isobutyl ketone	1.1
	17.1
	100.0

Component B	
Resin B	31.0
Xylene	45.9
Butyl Cellosolve	23.1
	100.0

Parts one and two of component A should be prepared separately and then blended. Components A and B should be mixed in the ratio 20 parts A : 1 part B by weight. The formulation contains 93% by weight of zinc dust in the dry film.

MARINE ZINC CHROMATE PRIMER

Component A	Parts by weight
Zinc chromate	17.4
Red oxide	10.9
Mica	5.5
Talc	13.2
Bentone 34	0.7
Epoxy resin (E.E.W. 500)	18.6
Methyl isobutyl ketone	6.2
Xylene	20.6
Butyl Cellosolve	6.9
	100.0

Component B	
Resin B	56.4
Xylene	43.6
	100.0

In the preparation of component A, the epoxy resin should first be dissolved and the other materials then incorporated. Components A and B should be mixed in the ratio 6 parts A : 1 part B by weight.

Zinc chromate must not be ground or stored in polyamide.

MARINE RED LEAD PRIMER

Component A	Parts by weight
Red lead	37.4
Red oxide	7.5
Talc	7.5
Mica	3.7
Bentone 34	0.5
Resin B	17.3
Xylene	21.6
Cellosolve	4.5
	100.0

Component B	
Epoxy resin (E.E.W. 500)	49.0
Methyl isobutyl ketone	16.4
Xylene	24.5
Cellosolve	10.1
	100.0

In the preparation of component A, the polyamide should first be dissolved and the other components then incorporated. Components A and B should be mixed in the ratio 1:40 parts A : 100 parts B by weight. If butyl cellosolve is used instead of Cellosolve, superior brushing properties will be obtained.

Alternatively a solution of the resin could be used and this is often a more cost effective formulation, avoiding the need to warm drums of viscous polyamide resin before compounding to assist their discharge.

HIGH BUILD RED LEAD COATING

Component A	Parts by weight
Red lead	30.0
Red oxide	3.0
Talc	13.0
Blanc fixe	12.0
Aerosil 300	1.0
Bentone 27	0.4
Epoxy resin (E.E.W. 200)	25.0
Tricresyl phosphite	3.0
Xylene	5.6
Cellosolve	7.0
	100.0

Component B	
Resin C	85.0
Xylene	15.0
	100.0

Component A and B should be mixed in the ratio 100 parts component A : 25 parts Component B.

MARINE COAL TAR ENAMEL

Component A	Parts by weight
Epoxy resin (E.E.W. 500)	30.2
Methyl isobutyl ketone	10.0
Butyl Cellosolve	6.5
Xylene	8.7
Coal Tar	44.6
	100.0

Component B	
Resin B	18.3
Xylene	59.7
Butyl Cellosolve	22.0
	100.0

In the preparation of Component A, the epoxy resin should first be dissolved in the solvent, and then blended with the coal tar. Components A and B should be blended in the ratio 100 parts A to 80 parts B.

SOLVENTLESS RED OXIDE BRUSHING PAINT

Component A	Parts by weight
Red iron oxide	27.7
Barytes	27.6
Epoxy resin (E.E.W. 200)	40.4
Phenol	4.3
	100.0

Component B	
Polyamino amide	77.8
Curing Agent K.54*	22.2
	————
	100.0

*Ex-Anchor Chemical Co. Ltd.

Components A and B should be mixed in the ratio 9 : 1 parts A : part B by weight.

WHITE GLOSS PAINT

Component A	Parts by weight
Titanium di-oxide (rutile)	34.67
Titanium di-oxide (anatase)	5.33
Epoxy resin (E.E.W. 500)	28.00
M.I.B.K.	13.33
Xylene	12.00
Cyclo hexanol	6.67
	————
	100.00

Component B	
Resin B (70% in xylene)	64.0
Xylene	36.0
	————
	100.0

Mix component A and B in the ratio 3:1 (A:B). The mixture has a pot life of 24 hours at ambient temperatures. The gloss paint is suitable for the following applications:

Superstructure of ships

Aircraft

Floors

Road Markings

Runway Markings

General Chemical and corrosion resistant paints.

A coat of primer (50 microns dry thickness) and gloss (48 microns dry thickness) were applied to a mild steel panel and allowed to cure for 24 hours at room temperature. The panel was cross hatched and exposed to salt spray testing (BS 3900) for 1200 hours. There was only very slight rusting on the scratches.

CLEAR SPRAYING LACQUER

Component A	Parts by weight
Epoxy resin (E.E.W. 500)	50.0
Methyl isobutyl ketone	25.0
Xylene	25.0
	100.0

Component B	
Resin B	50.0
Isopropanol	25.0
Toluene	25.0
	100.0

Mix components A and B in equal parts.

CLEAR SOLVENTLESS COATING
(Formulation 1)

Component A	Parts by weight
Epoxy resin (E.E.W. 200)	91.0
Phenol	9.0
	100.0

Component B	
Resin D	57.7
Tri (di-methyl amino methyl) phenol	42.3
	100.0

Components A and B should be mixed in the ratio 5.7 parts A : 1 part B by weight.

CLEAR SOLVENTLESS COATING
(Formulation 2)

Component A	Parts by weight
Epoxy resin (E.E.W. 200)	90.9
Cresylic acid*	9.1
	100.0

*Grade 50, ex-Midland Tar Distillers Ltd.

Component B	
Polyamino amide resin	76.1
Tri (di-methyl amino methyl) phenol	23.9
	100.0

Components A and B should be mixed in the ratio 4.4 parts A : 1 part B by weight.

WHITE SOLVENTLESS COATING

Component A	Parts by weight
Epoxy resin (E.E.W. 200)	57.2
Rutile titanium di-oxide	15.5
Barytes	16.1
Di-atomaceous earth	6.0
Thickening agent	1.2
Cresylic acid solution (50% in alcohol)	4.0
	100.0

Component B	
Resin D	100.0

Components A and B should be mixed in the ratio 2.5 parts A : 1 part B by weight.

SOLVENTLESS TAR COATING

Component A	Parts by weight
Epoxy resin (E.E.W. 200)	100.0

Component B	
Liquid tar	37.9
Resin C	31.9
Accelerator DMP30 (Rohm & Haas)	1.6
Thickening agent	2.8
Talc TX	25.8
	100.0

Components A and B should be mixed in the ratio 1 part A : 2.75 parts B.

SOLVENTLESS COATING FOR APPLICATION BY HEATED TWO COMPONENT AIRLESS SPRAY EQUIPMENT

Component A	Parts by weight
Epoxy resin (Epoxide equivalent 200)	65.4
Synthetic iron oxide	16.0
Asbestine 200μm	10.8
Microtal	5.3
Thickening agent	1.5
Silicone resin R281	1.0
	100.0

Component B	
Polyamino amide resin	100.0

Components A and B should be mixed in the ratio 1.5 parts A:1 part B.

CEMENTATIOUS SCREED

Component A	Parts by weight
Portland cement	15.0
Washed sand (No. 16 Sieve)	67.0
Liquid epoxy resin (Epoxide equivalent 200)	11.0
	93.0

Component B	
Wolfkur 1289	6.0

Component C	
Hardening accelerator (Tri-phenyl phosphite)	1.0
	100.0

Components A and B should be mixed, and then Component C added to give a total of 100 parts by weight. This represents a starting formulation for an epoxy modified cement suitable for trowel application, giving improved chemical resistance and resistance to dusting.

CLEAR LACQUER
For protection and decoration of woodwork under arduous conditions

Component A	Parts by weight
Araldite 6100	45.6
Xylene	9.8
MIBK	9.8
Butyl alcohol	4.6

Component B	
Versamid 115*	13.6
Xylene	5.8
Tri-methyl benzene	5.9
Butyl alcohol	4.9
	100.0

Mix components A and B in the ratio 2.3:1 shortly before use.

FLEXIBLE COATING
For use on flexible substrates
such as plastics and rubber

Component A	Parts by weight
Hypalon 30	16.75
Stearic acid	0.15
Epikote 1001	6.3
MEK	6.3
Toluene	50.5

Component B	
Versamid 100*	3.4
Butyl alcohol	1.7
Toluene	14.2
Silicone solution	0.7
	100.00

*VERSAMID ex-Cray Valley Products Ltd.

Mix components A and B in the ratio 4:1 shortly before use.

116

UNDERWATER COATING (Solvent free)

Component A	Parts by weight
Epikote 816	40.0
Silicone oil	1.0
Wetting agent	0.5
Thixotropic agent	1.0
TiO_2	10.0
Black iron oxide	2.0
Talc	12.5
Blank fix	23.0
2-EHGE	10.0
	100.0

Component B	
Epilink 149*	100.0

*EPILINK ex-AKZO Chemie.

Mix components A and B in ratio 3:1.

HIGH BUILD COATING

Component A	Parts by weight
Epikote 1001 x 75	32.0
Xylene	10.0
Wetting agent	0.7
Levelling agent	0.2
Thixotropic agent	1.5
Talc	36.0
TiO_2	10.0
Xylene	4.6
Ethyl glycol	5.0
	100.0

Component B	
Epilink 173×70*	100.0

*AKZO Chemie.

Mix components A and B in ratio 85:15.

Chapter VIII

Vinyl and
Acrylic Resins

120

Chapter VIII

Vinyl and
Acrylic Resins

The vinyl and acrylic group of resins is one of the most widely used in surface coating applications.

Vinyl and acrylic resins may be used on their own or in blends with other resins. Coating resins can be divided into two distinct categories:

  i) Thermoplastic polymer coatings.

  ii) Thermoset polymer coatings.

The thermoplastic types are long chain polymers with high molecular weight and film form without external chemical action. The thermosetting types are shorter-chain polymers containing reactive groups which can be 'cured' by application of heat or usually by reaction with another chemical type to form a cross-linked film.

Thermoplastic films harden by solvent evaporation, where as thermoset films harden by chemical reaction.

Vinyl resins are produced by the addition polymerisation of vinyl monomers. The term 'vinyl monomer' is applied to the species of molecules which contain a reactive $C=C$ double bond. Individual molecules (termed 'monomers') of a vinyl compound are capable of a reaction to form long chains of monomer units linked via $C-C$ bonds (termed a 'polymer'). The monomer units in the molecular chain may be exclusively of one species (a 'homopolymer') or the polymer may comprise of two or more species of monomer ('co-polymer', 'terpolymer', etc.).

VINYL AND ACRYLIC MONOMERS

A wide range of vinyl monomers are used to produce polymers for surface coating applications. They may be represented by the general formula:

$$CH_2 = CRX$$

Where R is usually a H atom or CH_3 group.

X however may be any one of a wide variety of species, including aromatic rings, halogens, amides, esters or substituted esters.

Examples of typical vinyl monomers are:

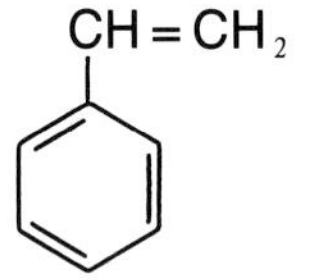

$$CH_2 = CHCl$$

$$CH_2 = CHO - \overset{\displaystyle ||}{\underset{\displaystyle O}{C}} - CH_3$$

Styrene   Vinyl chloride   Vinyl acetate

Of particular importance in surface coating applications are the two series of related monomers derived from: i) acrylic acid, and ii) methacrylic acid. These monomers are termed 'acrylates' and 'methacrylates' respectively, and polymers derived from either species are known collectively as acrylic resins. Typical acrylate and methacrylate monomers are shown below:

$$\underset{\displaystyle COOH}{\overset{\displaystyle CH_2 = CH}{|}}$$

$$\underset{\displaystyle COOH}{\overset{\displaystyle CH_2 = C - CH_3}{|}}$$

Acrylic acid   Methacrylic acid

Acrylic esters have the common structure:

$$\underset{\displaystyle CO_2R}{\overset{\displaystyle CH_2 = CH}{|}}$$

Whilst methacrylate esters are:

$$\underset{\displaystyle CO_2R}{\overset{\displaystyle CH_2 = CCH_3}{|}}$$

Examples are:

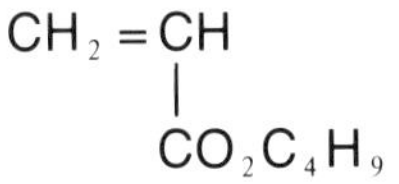

$$CH_2 = CH$$
$$|$$
$$CO_2C_4H_9$$

Butyl acrylate

$$CH_2 = CCH_3$$
$$|$$
$$CO_2CH_3$$

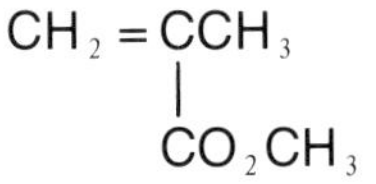

Methyl methacrylate

Various types of functionality can be built into acrylic or methacrylic monomers and thus be introduced into the final acrylic resin. The functional groups of vinyl (and acrylic) monomers which are commercially available and used to give a cross-linking capacity are hydroxyl, carboxyl, amide and epoxy.

Hydroxyl

Hydroxy functionality is usually provided by hydroxy acrylates, or alternatively, methacrylates which combine a double bond and a reactive hydroxyl group in the same molecule.

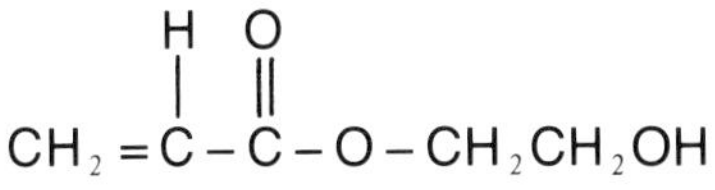

$$\text{H} \quad \text{O}$$
$$| \quad \parallel$$
$$CH_2 = C - C - O - CH_2CH_2OH$$

2-hydroxy ethyl acrylate

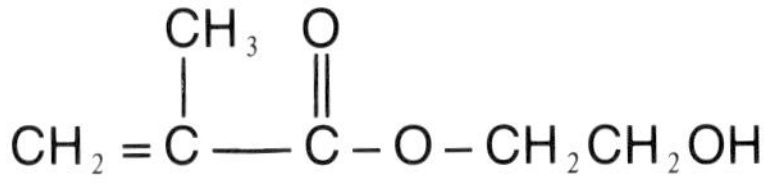

$$CH_3 \quad \text{O}$$
$$| \quad \parallel$$
$$CH_2 = C \longrightarrow C - O - CH_2CH_2OH$$

2-hydroxy ethyl methacrylate

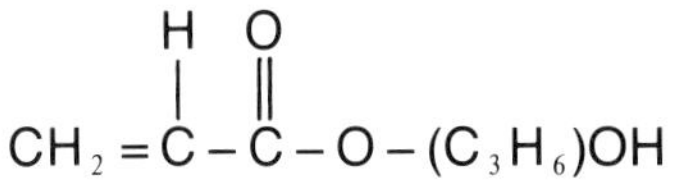

$$\text{H} \quad \text{O}$$
$$| \quad \parallel$$
$$CH_2 = C - C - O - (C_3H_6)OH$$

2-hydroxy propylacrylate

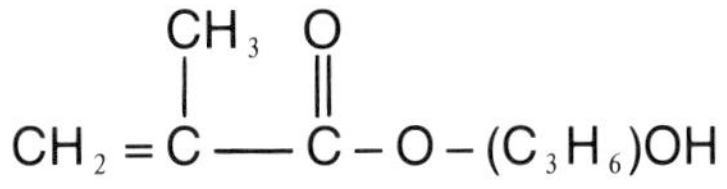

$$CH_3 \quad \text{O}$$
$$| \quad \parallel$$
$$CH_2 = C \longrightarrow C - O - (C_3H_6)OH$$

2-hydroxy propyl methacrylate

Carboxyl

The main carboxyl containing materials used to produce cross-linked acrylic polymers are acrylic or methacrylic acid. These have been discussed above.

Epoxy

Glycidyl acrylates or methacrylates are generally used to build epoxy groups into the polymer.

$$CH_2 = \overset{\overset{\displaystyle H}{|}}{C} - \overset{\overset{\displaystyle O}{\|}}{C} - O - CH_2 - CH - CH_2$$

$$CH_2 = \overset{\overset{\displaystyle CH_3}{|}}{C} - \overset{\overset{\displaystyle O}{\|}}{C} - O - CH_2 - CH - CH_2$$

Glycidyl acrylate   Glycidyl methacrylate

Glycidyl acrylate is volatile and toxic and finds limited application. Glycidyl methacrylate is less hazardous and is normally used in preference to glycidyl acrylate. Supplies of glycidyl acrylate are difficult to obtain outside of Japan.

Amide

This is another functional group regularly used in acrylic chemistry and is usually provided by acrylamide or methacrylamide.

$$CH_2 = \overset{\overset{\displaystyle H}{|}}{C} - \overset{\overset{\displaystyle O}{\|}}{\underset{\underset{\displaystyle NH_2}{|}}{C}}$$

$$CH_2 = \overset{\overset{\displaystyle CH_3}{|}}{C} - \overset{\overset{\displaystyle O}{\|}}{\underset{\underset{\displaystyle NH_2}{|}}{C}}$$

Acrylamide   Methacrylamide

Acrylamide is toxic and extreme care must be exercised during its use and handling. Disposal of empty containers must adhere to the relevant regulations.

Some monomers are often referred by initials such as M.M.A. for methyl methacrylate and V.A. for vinyl acetate. For compound names it is normal to use brackets, e.g. poly (methyl acrylate) indicates a polymer of methyl acrylate; polymethyl acrylate could be construed to mean an acrylate monomer with many methyl substituents.

The properties of some acrylic monomers along with a generalised outline of their preparations follow. This is for general interest.

Name	Mol. Weight	Typical Boiling Point °C	R.I.	Flash Point	S.G.
Styrene	104	146	1.5468	33°C closed cup	0.91
Acrylic acid	72	142 mpt 12°C	1.4224	57°C open cup	1.06
Methacrylic acid	86	168 mpt 15°C	1.4312	77°C closed cup	>1.0
Methyl acrylate	86	80	1.4003	−3°C open cup	0.950
Ethyl acrylate	100	99	1.4030	10°C open cup	0.917
Butyl acrylate	128	146	1.4160	38°C closed cup	0.984
Methyl methacrylate	100	101	1.412	11°C closed cup	0.939
Ethyl methacrylate	114	118	1.412	20°C open cup	0.910
Butyl methacrylate	142	163	1.422	54°C open cup	0.889
2-hydroxy ethyl acrylate	116	191	1.4505	99°C closed cup	1.107
2-hydroxy propyl acrylate	130	170	1.449	124°C open cup	1.054
2-hydroxy ethyl methacrylate	130	198	1.450	116°C open cup	1.07
2-hydroxy propyl methacrylate	144	197	1.448	107°C open cup	1.030
Acrylonitrile	53	77	1.3911	−1°C closed cup	0.80
Methacrylonitrile	67	90	1.4013		0.850
2-ethylhexyl acrylate	184	217	1.433	90°C open cup	0.880
Acrylamide	71	mpt 84°C			1.122
Methacrylamide	85	mpt 110°C			
Glycidyl acrylate	128	57 (2 mbar)		60°C	1.10
Glycidyl methacrylate	142	189		75°C open cup	

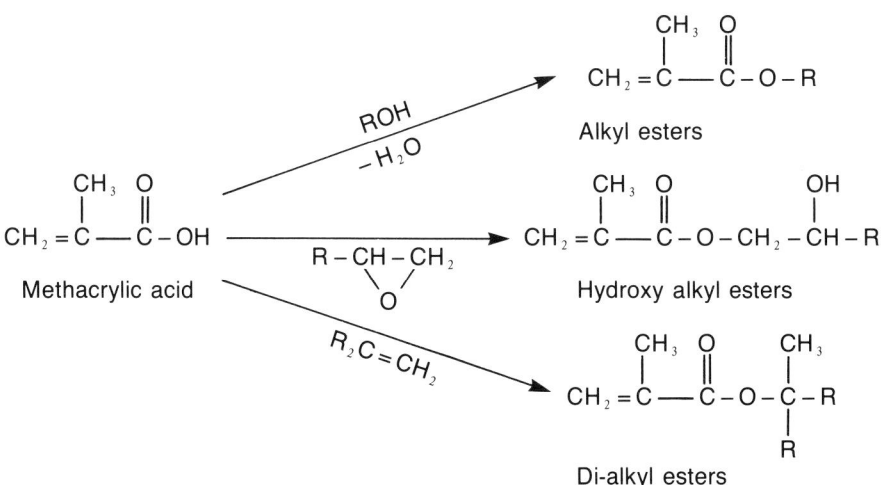

126

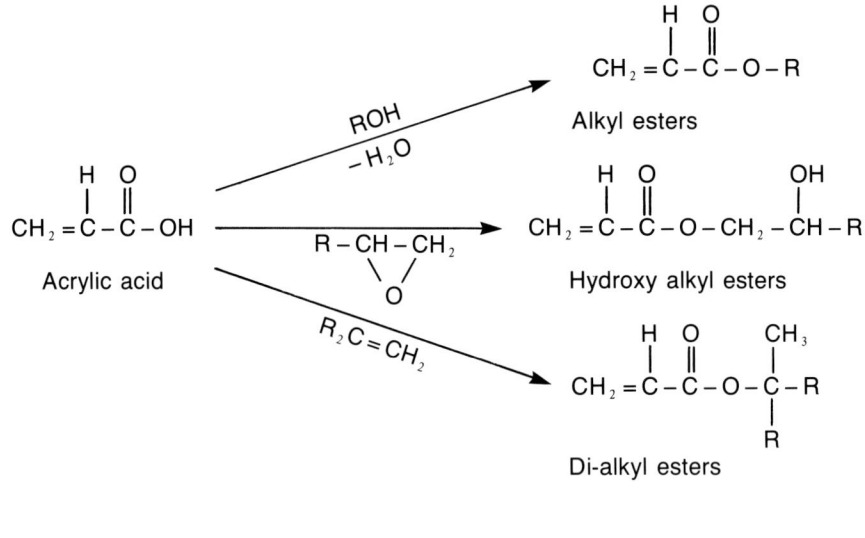

MONOMER SELECTION

There are a wide range of monomers available for use in the formulation of vinyl and acrylic resins. The selection of monomers is normally dictated by the performance, requirements and cost of the resulting product. As explained in Chapter 1, Volume I, few homopolymer resins are used in surface coatings. Combinations of monomers are used to give a compromise of their individual properties.

Homopolymers such as polyvinyl acetate are sometimes used for coating resins (e.g. emulsion paint) but improved properties can be obtained by co-polymerising more than one type of monomer to form a polymer such as poly vinyl acetate/versatate (the vera-static acid ester monomer acts as a plasticiser, flexibilising the otherwise brittle vinyl acetate film) or in polystyrene/butyl acrylate (the butyl acrylate adds flexibility and improves the reaction rate during manufacture of the polymer). Reactive polymers can be obtained by including in the polymer molecule monomeric units as hydroxyethyl acrylate or acrylamide which have functional pendant groups which can be used to cross-link the polymer to form a 'cured' coating.

The minimum film forming temperature and glass transition temperature are directly influenced by the proportion of a monomer in the polymer. For example, if the Tg of a styrene homopolymer is 100°C and of a butyl acrylate homopolymer is 54°C, then a copolymer formed from styrene and butyl acrylate will have a Tg between 100°C and 54°C, depending on the relative proportion of each monomer.

The following list and tables give general guidelines for monomer selection. Of course Tg, residual odour, viscosity and many other properties of the resulting polymeric film also have to be considered, when selecting mixtures of monomers. Some of these properties are discussed in the section on polymer properties.

Two other important factors which must be considered during the selection of monomers for industrial manufacture of typical surface coating resins are cost and toxicity.

The monomer combination used must not price the resin out of its intended market. Thus monomer selection may be a compromise of properties of the monomers unable to meet a target Raw Material Cost (R.M.C.). Obviously, small amounts of high cost monomers which may significantly upgrade performance can be incorporated in a low cost resin if the remainder of the ingredients are extremely low in cost. Thus in addition to technical requirements, the skilled resin chemist sometimes has to select different permulations of monomers to give an economic advantage.

It is also necessary to ensure that your production facility is capable of safe handling of any of the raw materials selected.

MONOMER	FILM PROPERTY
Methyl Methacrylate	Exterior durability Hardness Stain and water resistance
Styrene	Cost reduction Hardness
Butyl and higher acrylates	Flexibility Water resistance
Hydroxy (meth) acrylates	Functional groups for cross-linking
Acrylic and methacrylic acids	Functional groups Hardness
Vinyl Chloride	Cost reduction Durability Chemical resistance
Ethylene	Cost reduction Flexibility
Vinyl Acetate	Hardness
Vinyl Versatate	Flexibility

The exterior durability of acrylic homopolymers is illustrated in the table below:

Methyl acrylate	Poor	
Ethyl acrylate	Fair	
Butyl acrylate	Very good	Increasing durability ↓
Methyl methacrylate	Very good	
Ethyl methacrylate	Excellent	
Butyl methacrylate	Excellent	

Acrylic resins usually contain a considerable variety of monomers, the following table gives a general classification for monomers and helps to indicate which ones are most suitable for a particular application.

Monomer	Hardness	Flexibility	Resistance (Alkali)	UV retention Gloss retention	Solubility
α Methyl styrene			Excellent	Very poor	
Styrene			Excellent	Poor	
Vinyl toluene*	Increase in hardness (approx.) ↑	Increase in flexibility (approx.) ↓	Excellent	Poor	Increase in solubility (approx.) ↓
Acrylonitrile*			Fair/Poor	Fair/Poor	
Methyl methacrylate			Very good	Very good	
Ethyl methacrylate			Excellent	Excellent	
Butyl methacrylate			Excellent	Excellent	
Methyl acrylate			Very good	Poor	
Ethyl acrylate			Very good	Fair/Good	
Butyl acrylate			Very good	Very good	

The table demonstrates general trends, but anomalies exist. Two examples are vinyl toluene (good solubility) and acrylonitrile (very poor solubility).

Acrylic acid, methacrylic acid, acrylamide or methacrylamide tend to give films with good hardness, solvent and grease resistance.

The method of preparation of a polymer has a major influence on its properties. The addition polymerisation of the monomers may be carried out in bulk, suspension emulsion or solution. Surface coating resins are not normally prepared by bulk addition polymerisation (in contrast to condensation polymerisation resins) and suspension polymerisation is only rarely used for monomers other than vinyl chloride.

POLYMERISATION METHODS

Polymerisation involves the breaking of a carbon-carbon double bond. This is a highly exothermic reaction. Heats of reaction are of the order of 14-21 kilocalories per mole, depending on monomer type. This equates to about 1000 kilojoules per kilogram of monomer. The reaction temperature greatly influences the molecular weight and molecular weight distribution of the product. To obtain a consistent molecular weight product, it is necessary to accurately control the reaction temperature by removing the heat evolved during the reaction. Excess heat may be removed by use of a heat exchange medium, normally water, which is passed through coils set inside (or occasionally around the outside) of the polymerisation vessel.

Constant agitation of the reactants is necessary to aid in the dissipation of the heat generated during the reaction. As the molecular weight of the polymer increases so the viscosity of the reaction mass increases. Thus it becomes increasingly more difficult to agitate the mass effectively. Agitation can be improved by diluting the reactants with an organic solvent (or water) thus reducing the viscosity of the reaction mass. The diluent has the added effect of acting as a 'heat sink' for the heat of reaction, thus helping to dissipate the excess heat.

The example calculations below illustrate typical temperature rises experienced during copolymerisation and show the effect of the inclusion of a diluent (in this case water) acting as a heat sink, to modify the temperature increase arising from the heat of reaction.

Details of the heats of reaction for individual monomers are well documented in the literature.

Calculation of Temperature Rise during Polymerisation

Reactor Charge	Weight (gram)	Heat of Polymerisation (cal/mole)	Molecular Weight	Specific Heat
Ethyl acrylate	100	18600	100	0.47
Methyl methacrylate	50	13800	100	0.45

HEAT CAPACITY OF REACTOR CHARGE
(Weight × Specific Heat)

Ethyl Acrylate   100 × 0.47 =   47 cal/°C

Methyl Methacrylate   50 × 0.45 = 22.5 cal/°C

Total Heat capacity   = 69.5 cal/°C

HEAT EVOLVED DURING POLYMERISATION
(Number of Moles × Heat of Polymerisation)

Ethyl Acrylate	$\frac{100}{100}$ × 18600 = 18600 calories
Methyl Methacrylate	$\frac{50}{100}$ × 13800 = 6900 calories
Total heat evolved	= 25500 calories

Increase in temperature (ΔT) due to reaction exotherm:

$$\Delta T = \frac{\text{Total heat evolved}}{\text{Total heat capacity}} = \frac{25500}{69.5} = 366.9°C$$

Inclusion of an amount of water equivalent to the monomer charge results in the following:

Heat capacity of water: 1.0 × weight of water charged

		Heat capacity cal/°C
Water	150 g	150 × 1.0 = 150
Ethyl Acrylate	100 g	100 × 0.47 = 47
Methyl Methacrylate	50 g	50 × 0.5 = 22.5
Total heat capacity		= 219.5

The heat of polymerisation will remain unchanged at 25,500 calories.

Temperature increase

$$\Delta T = \frac{25500}{219.5} = 116.2°C$$

The dramatic decrease in the temperature increase illustrates the effectiveness of the use of water or solvent as a heat sink during addition polymerisation.

In the solution polymerisation process, where solvents are involved, a further reduction in temperature increase can be achieved by operating at the reflux temperature of the solvent. In this case the latent heat of vaporisation as well as the specific heat of the solvent can be utilised to remove heat from the reactor.

Thus different techniques of polymerisation have been partially devised to overcome the problem of heat removal during polymerisation.

Five methods of free radical polymerisation are used commercially. A brief description of each of these methods shows some of the advantages and difficulties of each method. Each is discussed in more detail under their respective sections.

BULK POLYMERISATION

This involves heating monomer and initiator, possibly with the addition of a transfer agent. The problem of heat transfer through a large bulk of viscous liquid or gel results in poor cooling. Consequently, very long reaction times are required to ensure that there is no overheating. In practice bulk polymerisation is carried out in completely filled disposable containers held either in an oven or a water bath to control temperature. After polymerisation the container is cut away and the product is crushed. This restricts this method of production to polymers which are brittle at room temperature. Bulk polymers have minimum contamination but reaction control is poor, so reproducibility is bad. This method is seldom used for producing surface coating polymers. It is normally confined to the manufacture of high molecular weight polymers for the packaging and moulding industries.

SUSPENSION POLYMERISATION

A monomer solution of initiator is dispersed in water and heated. It is a bulk polymerisation in which relatively small droplets of bulk monomer are polymerised and cooled by water. The relatively low viscosity of the dispersion allows efficient cooling. The stability of the dispersion presents problems both during polymerisation where coagulation must be avoided as well as in the final dispersion where end use may involve high shear stresses. There is some disagreement about the droplet size in suspension polymerisations with values ranging from 0.5 to 5000 microns, but many commercial products fall into the $5-20$ micron range. As with bulk polymerisation the Tromsdorf effect gives a polymer heterodisperse in molecular weight. Reproducibility is good and fast reaction times can be achieved, although it is not easy to get conversion above 98%. The suspension method is seldom used for surface coatings. Some solid acrylic ink resins (for solvent based inks) are manufactured by a suspension technique.

SOLUTION POLYMERISATION

This provides two methods of cooling; the lower viscosity allows some jacket cooling and by carrying out the reaction under reflux the latent heat of vaporisation leads to very efficient cooling. The reflux operation also provides an effective temperature control for the polymerisation.

Solution polymers have low molecular weights and high polydispersity. The absence of additives aids stability to light and ageing and also enhances gloss. Disadvantages include difficulty in obtaining high conversion, cost, possible fire and health hazards of the solvents and lack of flexibility in formulation.

EMULSION POLYMERISATION

This differs from the suspension method in using water soluble initiators. The advantages are good reproducibility, rapid reaction, high conversion and low cost. Disadvantages are the presence of surfactants or colloids, salts, etc., which detract from optimum film performance and poor film integrity. Emulsion polymers have higher molecular weights than those obtained by any other free radical process. They also allow greater flexibility in the preparation of copolymers.

NON-AQUEOUS DISPERSION (N.A.D.'s)

Monomers are polymerised in the presence of organic liquids which are non-solvents for the polymers. Stabilisation is different to the aqueous suspension technique and generally the product is sold as a liquid, unlike suspension polymers which are washed and dried, and sold as solids. N.A.D.'s offer high solids, low viscosity coatings.

The following tables compare in an oversimplified manner the different polymerisation techniques mentioned:

Comparison of the Major Components of Different Addition Polymerisation Techniques

This table is an oversimplified comparison of the major components present for each type of polymerisation. Additives have been excluded.

TECHNIQUE	COMPONENTS
Bulk	Monomer(s) + initiator(s)
Solution	Monomer(s) + initiator(s) + solvent(s)
Suspension	Monomer(s) + initiator(s) + water + suspending agents
N.A.D.	Monomer(s) + initiator(s) + organic liquids + stabiliser(s)
Emulsion	Monomer(s) + initiator(s) + water + (colloid(s)) + (buffer(s)) + (surfactant(s))

Comparisons of some Properties of Resins resulting from Different Addition Polymerisation Techniques

The following is an over generalised comparison on the properties of the resins resulting from different techniques of polymerisation (N.A.D.'s omitted). In addition, some of the constraints of processing are compared.

PROPERTY	HIGH	LOW
Molecular weight	Emulsion >>	Solution, suspension, bulk
Purity of resulting resin	Bulk > Solution >	Suspension > Emulsion
Practical limits to non-volatile content during processing	Bulk (100%) > Solution (60–70%) > Emulsion (50–60%) > Suspension (30–35%)	
Instability during processing	Suspension > Emulsion >>	Solution Bulk
Heat removal problems	Bulk >> Solution >	Emulsion suspension

Typical Characteristics of
Solution and Emulsion Polymer

The typical characteristics of solution and emulsion polymers are compared in the following table:

PROPERTY	SOLUTION	EMULSION
Molecular weight	10,000 to 50,000	100,000 to 1 million
Viscosity	Very dependent on molecular weight	Low and independent of molecular weight
Rheological properties	Newtonian	Pseudoplastic
Solubility of Film	Soluble before cross-linking	Not redispersible

The majority of vinyl and acrylic surface coating resins are manufactured by either solution or emulsion polymerisation techniques.

THEORETICAL CONSIDERATIONS

Chapter 1 of Volume I contains a detailed review of the kinetics of free radical addition polymerisation and other theoretical properties. In this chapter, the application of theory to practice is considered with examples of materials actually used. Where some of the concepts are specific to a particular method of polymerisation the points are discussed in the appropriate section.

Mechanism of Polymerisation

Vinyl and acrylic monomers undergo addition polymerisation to form long chains, viz:

$$CH_2 = CRX \xrightarrow{\text{Initiator}} -(CH_2 - CRX)-$$

$$\text{Monomer} \qquad\qquad\qquad\qquad\qquad \text{Polymer}$$

Unlike condensation polymerisation, addition polymerisation leads to the production of high molecular weight polymers. There is also no release of water or other small molecules characteristic of condensation reactions. In a rigorous treatment, polyurethanes would be classified as addition polymers but the discussion in this chapter is restricted to addition polymerisation through unsaturated carbon bonds.

Polymerisation can be considered to occur in three distinct and separate stages, viz:

i) **Initiation:** Start of growth of the polymer chain by the attack of an initiation species on the double bond of the vinyl compound.

ii) **Propagation:** Growth of the polymer chain by successive additions of monomer units.

iii) **Termination:** Deactivation of the polymer chain terminating the further addition of monomer units.

Initiation

Initiation is brought about by one of three types of initiation:

a) **Free Radical Initiation**

$$RH \longrightarrow R^\bullet \quad + \quad CH_2 = CHX \longrightarrow RCH_2 - \overset{\bullet}{C}H_2$$

Initiator   Free radical   Monomer   Initiated species

b) **Anionic Initiation**

$$B^+A^- + CH_2 = CHX \longrightarrow A - CH_2 - CHX^-B^+$$

Radical ion

c) **Cationic Initiation**

$$A^-B^+ + CH_2 = CHX \longrightarrow B - CH_2 - CHX^+A^-$$

Most vinyl monomers will undergo initiation by either type of initiation. Ionic initiation is rarely used for polymers for surface coating applications. They are normally restricted to the production of the so called 'high polymers' which are employed in packaging, moulding and structural use areas, e.g. polyethylene, polypropylene, polyvinyl chloride, etc.

However, for certain combinations of monomers the only method of polymerisation is by an ionic technique. Some of the polyethers would fall into this category.

Initiator molecules can decompose in a variety of ways. Some molecules decompose to give two free radicals, e.g.:

$$R - \underset{\underset{O}{\|}}{C} - O - O - \underset{\underset{O}{\|}}{C} - R \quad \xrightarrow{\text{Heat}} \quad 2R - \underset{\underset{O}{\|}}{C} - O^\bullet$$

$$2R - \underset{\underset{O}{\|}}{C} - O^\bullet \quad \xrightarrow{\text{Heat}} \quad 2R^\bullet + 2CO_2$$

With other types the resulting free radicals can sometimes undergo further reaction with even the elimination of other molecules, e.g.

$$R-\underset{\underset{O}{\|}}{C}-O-O-\underset{\underset{CH_3}{|}}{\overset{\overset{CH_3}{|}}{C}}-CH_3 \quad \xrightarrow{\text{Heat}} \quad R-\underset{\underset{O}{\|}}{C}-O^\bullet \; + \; CH_3-\underset{\underset{CH_3}{|}}{\overset{\overset{CH_3}{|}}{C}}-O^\bullet$$

$$R-\underset{\underset{O}{\|}}{C}-O^\bullet \quad \xrightarrow{\text{Heat}} \quad R^\bullet \; + \; CO_2$$

$$CH_3-\underset{\underset{CH_3}{|}}{\overset{\overset{CH_3}{|}}{C}}-O^\bullet \quad \xrightarrow{\text{Heat}} \quad \overset{\bullet}{C}H_3 \; + \; CH_3-\underset{\underset{O}{\|}}{C}-CH_3$$

There are other mechanisms of initiation fragmentation but they are outside the scope of this chapter. It is the generation of free radicals rather than the scission process which is important here.

The most commonly used initiators are of the free radical type. Free radical initiators are substances which decompose to give short life time highly reactive species termed 'free radicals'. Examples of this type of initiator are peroxides and diazo compounds.

Decomposition may be brought about by heating the initiator (thermal decomposition) or under the influence of another chemical species forming an oxidation-reduction couple, commonly termed Redox initiation.

THERMAL INITIATION

The thermal decomposition of peroxides and azo compounds can be represented as follows:

Benzoyl peroxide $\xrightarrow{\text{Heat}}$ 2 Benzoate radical (free radical)

The benzoate radical may decompose further to give phenyl radicals, viz:

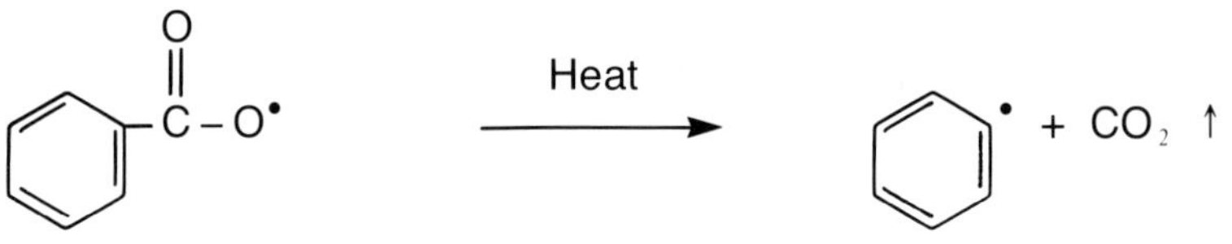

The ratio of benzoate radicals to phenyl radicals depends upon the reaction conditions. Another example of thermal decomposition is that of AZDN (Azo bis isobutyronitrile):

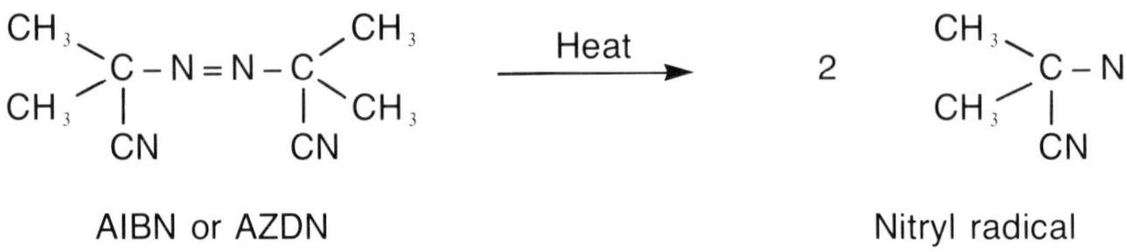

AIBN or AZDN

Nitryl radical

Further decomposition of the nitryl radical can and frequently does occur forming nitrogen radicals which may combine to eliminate gaseous nitrogen.

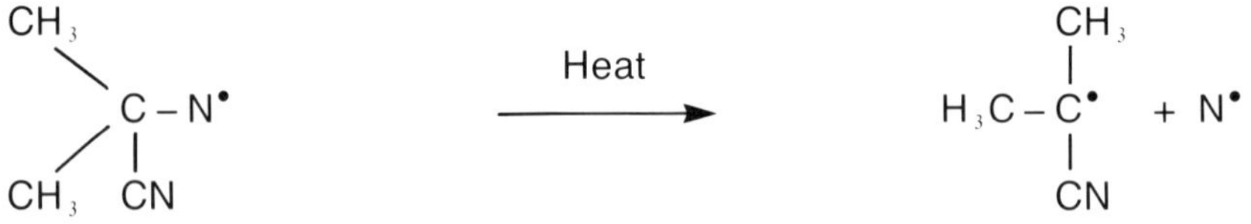

$$2 \ N^\bullet \longrightarrow N_2 \uparrow$$

The use of AZDN in suspension polymerisation can lead to the formation of a proportion of 'blown beads' or 'floaters'. These particles contain trapped nitrogen. This creates voids in the spherical particles and lowers the density from greater than one for a solid bead to less than one for a blown one, with the resulting bead floating rather than sinking in the suspension medium.

Thermal initiators require careful handling and warehousing. Some require refrigerated storage while others are quite stable at ambient temperatures. Suppliers recommendations must be observed when storing or using these materials.

REDOX INITIATION

Free radicals can be produced by thermal or chemical reactions. The latter are classed as Redox initiators because a reducing agent catalyses the decomposition of a peroxide compound forming a Redox couple. (Reduction−oxidation couple.)

Generation of free radicals by a Redox mechanism can occur at relatively low temperatures (even below ambient). This has many advantages for aqueous emulsion polymerisation. For gaseous monomers (e.g. vinyl chloride, ethylene) solubility and hence amount copolymerised with other vinyl monomers (e.g. vinyl acetate) depends upon the temperature and pressure. As a general rule, the lower the temperature and higher the pressure the greater the solubility. For some polymer latices (emulsions) thermal destabilisation may occur during polymerisation, thus Redox rather than thermal initiation is used.

Redox couples are extremely sensitive to impurities, particularly metal ions, unlike thermally initiated systems which tend to be more tolerant of conditions.

The generation of radicals for Redox and thermal initiation can be on different time scales. As a generalisation, for polymerisation thermal initiators generally retain their 'initiating potential' for a significant period (often measured in hours) after their addition (see half life section) unlike Redox initiators which are instantaneous (minutes rather than hours), with respect to the addition of one of the initiators. There are two distinct methods of adding Redox initiators:

i) Continuous addition of both components.

ii) Addition of one component to reactor initially and continuous addition of the other.

These two classifications give rise to the terminology of fast and slow Redox initiator systems.

The other common use for Redox initiation is towards the end of polymerisation where Redox systems are used for 'scavenging' (polymerising) residual monomers.

Initiators can be oil or water soluble or both, enabling initiation at the oil-water interface.

Redox initiation is covered in detail in the emulsion polymerisation section. However, two typical Redox systems are:

$$R-O-O-H + Fe^{2+} \longrightarrow RO^{\bullet} + OH- + Fe^{3+}$$

$$\text{Hydro peroxide} \qquad\qquad\qquad \text{Radical}$$

Multi-component Redox systems are used in emulsion polymerisation which can generate both free radicals and radical ions, viz:

$$S_2O_8{}^{2-} + Fe^{2+} \longrightarrow {}^{\bullet}SO_4- + SO_4{}^{2-} + Fe^{3+}$$

$$\text{Persulphate ion} \qquad\qquad\qquad \text{Radical ion}$$

Structure of Some Initiators

Structures of some typical peroxide or catalysts for the formation of free radicals are included for reference purposes.

Di-benzoyl peroxide
(Diacyl peroxides)

$$\text{Ph}-\overset{\overset{\displaystyle O}{\|}}{C}-O-O-\overset{\overset{\displaystyle O}{\|}}{C}-\text{Ph}$$

Azo-bis-isobutyronitrile
(AZO initiators)

$$CH_3-\underset{\underset{\displaystyle CN}{|}}{\overset{\overset{\displaystyle CH_3}{|}}{C}}-N=N-\underset{\underset{\displaystyle CN}{|}}{\overset{\overset{\displaystyle CH_3}{|}}{C}}-CH_3$$

t-butyl peroxybenzoate
(Peresters)

$$CH_3-\underset{\underset{\displaystyle CH_3}{|}}{\overset{\overset{\displaystyle CH_3}{|}}{C}}-O-O-\overset{\overset{\displaystyle O}{\|}}{C}-\text{Ph}$$

t-butyl hydroperoxide
(Alkyl hydroperoxides)

$$CH_3-\underset{\underset{\displaystyle CH_3}{|}}{\overset{\overset{\displaystyle CH_3}{|}}{C}}-O-OH$$

di-t-butyl peroxide
(Dialkyl peroxides)

$$CH_3-\underset{\underset{\displaystyle CH_3}{|}}{\overset{\overset{\displaystyle CH_3}{|}}{C}}-O-O-\underset{\underset{\displaystyle CH_3}{|}}{\overset{\overset{\displaystyle CH_3}{|}}{C}}-CH_3$$

Perdicarbonates

$$R-O-\overset{\overset{\displaystyle O}{\|}}{C}-O-O-\overset{\overset{\displaystyle O}{\|}}{C}-O-R$$

Perketals

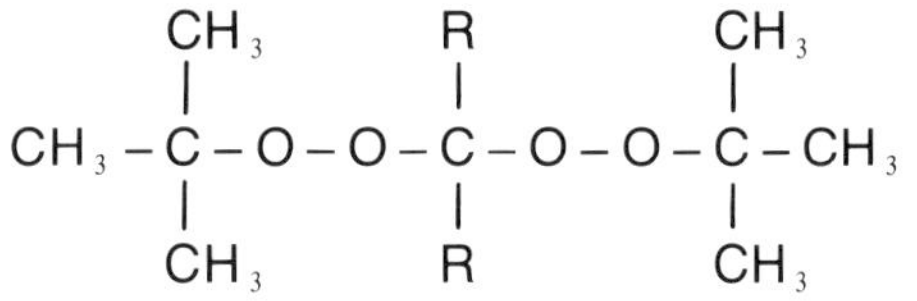

$$CH_3-\underset{\underset{CH_3}{|}}{\overset{\overset{CH_3}{|}}{C}}-O-O-\underset{\underset{R}{|}}{\overset{\overset{R}{|}}{C}}-O-O-\underset{\underset{CH_3}{|}}{\overset{\overset{CH_3}{|}}{C}}-CH_3$$

Ketone peroxides
(various isomers)

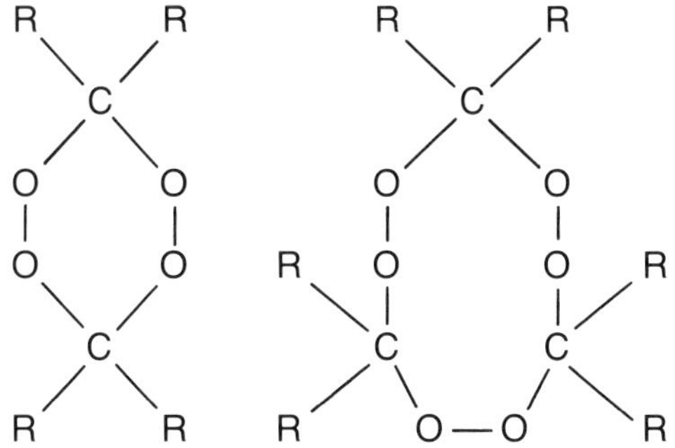

Undesirable Decomposition of Peroxides

Accelerated decomposition of organic peroxides, particularly in bulk, can present a major hazard. It is usually caused by:

Heat

Reducing agents

Strong alkalis or acids

Metallic contaminants *(see Chapter on Unsaturated Polyesters).*

HALF LIVES OF INITIATORS

Under the influence of heat, thermal initiators like peroxy, compounds decompose to form free radicals. Polymerisation can only proceed efficiently and economically, if sufficient free radicals are present in a given unit of time. Too many free radicals can have deliterious effects upon the properties of the resulting resin including excessive grafting, too low molecular weight or high degree of oxidation or polar chain ends. Thus, it is essential to know how the number of free radicals relate to the initiator, temperature and conditions used.

The relationship normally used to determine this is the initiator half life ($t\frac{1}{2}$). Only thermal initiation will be considered here. Classical derivation of half life gives $t\frac{1}{2}=\ln2/k$.

But for initiators k is temperature dependent, and is related to the activation energy (of decomposition) (E), the gas constant (R) and temperature in degrees absolute (T). This is the standard Arrhenius equation.

A maximum velocity constant is also incorporated (k_{max})

$$k = k_{max}e^{-E/RT}$$

Thus $t_{\frac{1}{2}} = \ln2/(k_{max}e^{-E/RT})$

$$\ln t_{\frac{1}{2}} = \ln(\ln2/k_{max}) + E/RT$$

The term $\ln(\ln2/k_{max})$ can be considered constant. Thus plotting the Naperian (natural) logarithm of $t_{\frac{1}{2}}$ against $1/T$ should be linear with a gradient of $E/(2.3R)$.

Organic peroxides have activation energies in the region $100-150$ kJ/molecules (Interex booklet, P321, Organic Peroxides).

Peroxides with low energies of activation have moderately linear rates of decomposition with respect to temperature, unlike the higher energies which give a large increase in rate of decomposition (decreased $t_{\frac{1}{2}}$) for a small increase in temperature. The former products are suitable for non-isothermal reactions.

The kinetics indicate that $t_{\frac{1}{2}}$ obeys first order kinetics. As will become apparent free radical decomposition normally follows non-ideal kinetics due to the influence of factors other than temperature affecting the rate of decomposition. The order of decomposition is between first and second order.

The stability of an initiator at any given temperature is measured in terms of its half life. Comprehensive details of half lives over a wide range of temperatures are available from the literature, particularly that of initiator suppliers. This detailed information is essential when formulating vinyl and acrylic polymers for surface coating applications. Half lives ($t_{\frac{1}{2}}$) for some common free radical initiators which have been taken from the literature are shown in the following table to illustrate the effect of the temperature on the decomposition rate.

HALF LIVES FOR SOME COMMON FREE RADICAL INITIATORS AT VARIOUS TEMPERATURES

Initiator	Temperature	$t_{\frac{1}{2}}$ (half life)	Optimum Temperature Range
Di-cumyl Peroxide	100°C	20 hours	130 – 140°C
	120°C	5.5 hours	
	130°C	2 hours	
	140°C	35 minutes	
	150°C	12 minutes	
	160°C	4.5 minutes	

HALF LIVES FOR SOME COMMON FREE
RADICAL INITIATORS AT VARIOUS TEMPERATURES (cont.)

Initiator	Temperature	$t\frac{1}{2}$ (half life)	Optimum Temperature Range
Di-tertiary Butyl Peroxide	130°C	6 hours	140 – 150°C
	140°C	2 hours	
	150°C	40 minutes	
	160°C	15 minutes	
Tertiary Butyl Perbenzoate	110°C	5.5 hours	115 – 130°C
	120°C	1.75 hours	
	130°C	35 minutes	
	140°C	12 minutes	
	150°C	4.5 minutes	
Tertiary Butyl Perpivalate	60°C	6 hours	70 – 80°C
	70°C	1.25 hours	
	80°C	20 minutes	
	90°C	9 minutes	
Di-benzoyl Peroxide	80°C	4 hours	90 – 100°C
	90°C	1.25 hours	
	100°C	25 minutes	
	110°C	8.5 minutes	
Azobiz Isobutryl Nitrile	64°C	10 hours	75 – 90°C
	82°C	60 minutes	
	100°C	6 minutes	
	120°C	1 minute	

However, half life for any given initiator molecule doesn't only depend upon the temperature, it also depends upon its environment and concentration.

As an example, di-benzoyl peroxide has a half life of 85 minutes at 90°C in styrene polymerisation and 65 minutes at 90°C when used as a polymer cross linking agent (concentrations may differ).

Normally half lives are quoted for 0.1 molar solutions in benzene or other non-polar, suitable solvent. Under these conditions half life is approximately 75 minutes.

Half life times are affected by the concentration of peroxide used, the lower the concentration the longer the half life time.

142

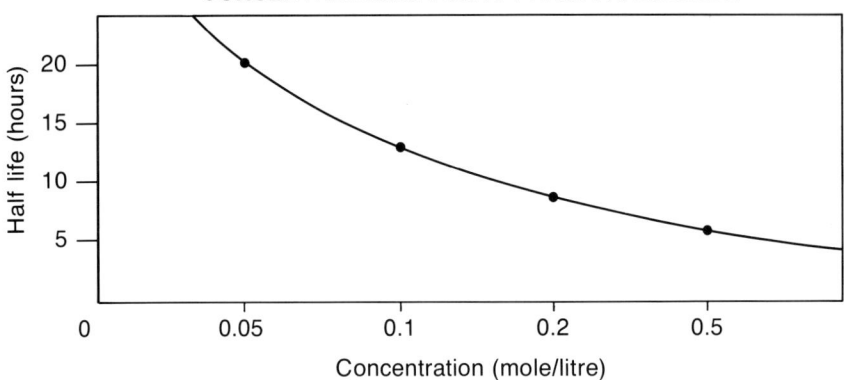

Some peroxides can be attacked and directly split by certain solvents. Primary radicals formed during decomposition can form secondary radicals with some solvents or indeed other peroxide molecules. These secondary radicals can then cause decomposition of other peroxide molecules. This is often termed 'induced decomposition'. As a general rule, half lives increase in solvents in the following order: alcohols, ethers, aliphatic hydro carbons, aromatics, highly halogenated solvents, for the same concentration for a given initiator.

In tri-chloro ethylene primary radicals react with solvent and only unreactive secondary radicals are formed. In this case $t_{\frac{1}{2}}$ is significantly larger than if the peroxide were in benzene. Factors of two to three times longer are not uncommon.

In isododecane, $t_{\frac{1}{2}}$ for t butyl perpivalate is up to three times longer than if it were in acetophenone.

Induced decomposition reduces the half life from a first order to between first and second order reaction.

Increasing pressure tends to increase $t_{\frac{1}{2}}$. Under normal polymerisation conditions this is irrelevant because the effects are much less than those of changing temperature and the pressure changes are small (less than one bar).

However, for polymerisation of gaseous monomers (e.g. ethylene copolymers) high pressures are used and the effect of pressure on $t_{\frac{1}{2}}$ can be important. Examples from the literature (Interex booklet, P321) show $t_{\frac{1}{2}}$ increasing from 20 seconds at 1 bar to 28 seconds at 1000 bar to 42 seconds at 3000 bar for t-butyl perpivalate. Very few surface coating resins are polymerised at pressures above 100 bar. Impurities and other factors can have a greater influence on $t_{\frac{1}{2}}$ than pressure. Certain metal ions can have devastating effects. This is sometimes used to advantage in redox polymerisation.

It should be stressed that the figures quoted are for guidance and comparison only. It is necessary for the resin chemist to experimentally determine the best initiator type and concentration for their particular system. External constraints like the solvent system and hence maximum processing temperature, tolerance for extent of grafting and ease of addition of solid, paste or liquid initiator on the plant all influence the choice of initiator.

A half life of 30 minutes to two hours is preferred for both commercial (and academic work). Azo isobutyro nitrile, abbreviated to AIBN or AZDN does not cause oxidation which can occur with peroxide initiation, particularly where liable hydrogen atoms (e.g. acrylate monomers) are present. As a generalisation a greater degree of grafting occurs with a peroxide initiated system compared to an AZDN one (for similar concentrations, conditions and monomers).

In a latter section, the determination of theoretical versus actual conversion during polymerisation is discussed. It is strongly recommended that this procedure is used for the system and temperature to be polymerised on the laboratory bench before undertaking pilot or larger scale manufacture. Build up of unreacted monomers must be avoided. Careful selection of initiator and processing conditions can reduce the possibilities of this happening.

The free radicals produced by the decomposition of the initiator attacks the bond of a monomer to produce a monomer radical.

Di-cumyl peroxide   Di-cumyl radical

Methyl   Initiated
methacrylate   monomer radical

The number of initiated chains present at the initiation stage of polymerisation can be estimated from the following:

$$R_1 = 2fk_d [I]$$

Where R_1 = rate of initiation
k_d = rate coefficient for the decomposition of initiator to give free radicals.

[I] = initiator concentration   f = initiator efficiency

At time t the number of incorporated initiation fragments (and hence the number of growing polymer chains) is given by the equation:

$$\text{Number of incorporated initiation fragments} = 2fk_d \, [I] \, t$$

The temperatures used for conventional emulsion or aqueous polymerisation excludes the use of the majority of thermal peroxide initiators because of their excessively long half lives at reaction temperatures.

Propagation

Once the initiated monomer radical is formed, propagation occurs rapidly with monomer units adding successively onto the growing polymer chain with the free radical being retained on the vinyl C atom of the end monomer unit.

$$RO-CH_2-\overset{\bullet}{C}HX + CH_2 = CHX \longrightarrow RO-CH_2-CHX-CH_2-\overset{\bullet}{C}HX$$

$$\downarrow nCH_2 = CHX$$

$$RO-CH_2-CHX-(CH_2-CHX)_n-CH_2-\overset{\bullet}{C}HX$$

The length of the growing polymer chain has no effect on the reaction rate and each propagation step proceeds in accordance with the propagation constant kp for that monomer at that temperature. The factors governing kp are the ease with which the monomer forms a free radical and the reactivity of the monomer radical once formed. In general monomers that form radicals easily tend to polymerise more slowly than monomers that do not form radicals readily. For example, styrene monomer readily forms a radical monomer.

$$CH = CH_2 \qquad + \qquad RO^{\bullet} \longrightarrow \qquad RO-CH-\overset{\bullet}{C}H_2$$

However, the styrene monomer radical is stabilised by de-localisation of the radical in the π electron cloud of the benzene ring leading to a relatively slow rate of propagation when forming polystyrene homopolymer.

Vinyl acetate, however, does not readily form monomer radicals at temperatures below c.a. 50°C, but once formed the monomer radical is unstable and is thus highly reactive and propagation to form polyvinyl acetate homopolymer is extremely rapid.

For a homopolymer reaction the rate of propagation is given by the following equation:

$$\text{Rate of propagation (Rp)} = kp \, [M] \, [M^\bullet]$$

Where kp = rate constant for propagation for monomer M

$[M]$ = concentration of monomer M
$[M^\bullet]$ = concentration of monomer radicals

In the case of a copolymerisation reaction involving the monomers (A and B) there are four possible rate constants, i.e. one for each of the possible propagation steps. However, the rate of propagation can still be related to the propagation constant and the monomer and monomer radical concentrations as shown above.

The four possible propagation steps are:

i) $\sim\!\!\sim A^\bullet + A \longrightarrow \sim\!\!\sim A - A^\bullet$ where $kp = kp \,(AA)$
ii) $\sim\!\!\sim A^\bullet + B \longrightarrow \sim\!\!\sim A - B^\bullet$ where $kp = kp \,(AB)$
iii) $\sim\!\!\sim B^\bullet + A \longrightarrow \sim\!\!\sim B - A^\bullet$ where $kp = kp \,(BA)$
iv) $\sim\!\!\sim B^\bullet + B \longrightarrow \sim\!\!\sim B - B^\bullet$ where $kp = kp \,(BB)$

COPOLYMERISATION

In Chapter 1, Volume I, the theory of addition copolymerisation was discussed. To summarise a propagating chain Mi can add different types of monomer Mj. If we restrict the discussion to two monomers, M_1 and M_2 then four reactions are possible.

$$\sim\!\!\sim M_1^\bullet + M_1 \longrightarrow M_1 \, M_1^\bullet \qquad \text{rate constant } k_{1,1}$$
$$\sim\!\!\sim M_1^\bullet + M_2 \longrightarrow M_1 \, M_2^\bullet \qquad \text{rate constant } k_{1,2}$$
$$\sim\!\!\sim M_2^\bullet + M \longrightarrow M_2 \, M_1^\bullet \qquad \text{rate constant } k_{2,1}$$
$$\sim\!\!\sim M_2^\bullet + M_2 \longrightarrow M_2 \, M_2^\bullet \qquad \text{rate constant } k_{2,2}$$

Reactivity ratios r_1 and r_2 were derived to relate to these rate constants, viz:

$$r_1 = k_{1,1}/k_{1,2} \qquad\qquad \text{and} \qquad\qquad r_2 = k_{2,2}/k_{2,1}$$

Some typical reactivity ratios are shown below:

SOME REACTIVITY RATIOS

Monomer 1 →	Acrylonitrile		M.M.A.		Styrene		Vinyl acetate	
Monomer 2 ↓	r_1	r_2	r_1	r_2	r_1	r_2	r_1	r_2
Acrylonitrile	1	1	1.3	0.12	0.40	0.045	0.05	4.2
Acrylamide	0.8	1.3	2.5	0.8	1.13	0.59	—	—
Acrylic acid	0.13	6.0	2.3	0.3	0.25	0.14	0.03	5.0
Butadiene	0.03	0.18	0.3	0.7	0.5	1.4	—	—
Ethyl acrylate	1.15	0.67	2.0	0.26	0.80	0.20	—	—
Ethylene	—	—	17	0.2	—	—	1.1	1.2
1 – Hexene	12.2	0	—	—	0.19	9.8	—	—
Maleic anhydride	6	0	3.8	0.02	0.04	0.01	0.06	0.01
M.M.A.	0.15	1.2	1	1	0.50	0.50	0.025	20
Styrene	0.045	0.40	0.50	0.50	1	1	0.01	50
Vinyl acetate	4.2	0.05	20	0.025	50	0.01	1	1
Vinylidine chloride	0.65	0.03	2.5	0.4	2.0	0.14	0.05	5.0

When both r_1 and r_2 are low, e.g. for maleic anhydride with either styrene or vinyl acetate, each radical will react preferentially with the other species and an alternating copolymer will be formed. When r_1 and r_2 are both nearly unity, e.g. acrylonitrile and acrylamide, a random copolymer will be produced. If both r_1 and r_2 are high a block copolymer would be formed; no such values have been reported for free radical polymerisation. When one ratio is high and the other low, initial copolymer will be rich in the monomer with the high ratio; as polymerisation proceeds this monomer will be depleted and the copolymer composition will become increasingly rich in the other monomer. It is possible to obtain a random copolymer by adding a small quantity of monomers, rich in the less reactive monomer, and then adding the balance of the monomer mixture at a rate equal to that of polymerisation. The reactivity ratios can be summarised as follows:

1. If $r_1 > 1$ then $M_1^•$ prefers to add M_1.
2. If $r_1 < 1$ then $M_1^•$ prefers to add M_2.
3. If $r_1 = 1/r_2$ then the probability of $M_1^•$ adding either M_1 or M_2 is equal and a perfectly random copolymer is formed.
4. If $r_1 = r_2 = 0$ then the two polymers do not homopolymerise but will copolymerise (provided $k_{1,2}$ and $k_{2,1}$ are both > 0) to form a perfectly alternating copolymer.
5. If $r_1 > 0$ and $r_2 = 0$ then $M_2^•$ does not add M_2 but M_1 (provided $k_{2,1} > 0$) to form $M_1^•$ which then adds either M_1 or M_2 depending upon the magnitude of r_1.

For the majority of monomers the values of r_1 and r_2 are such that there is uneven depletion of one of the species, resulting in the composition of a copolymer molecule formed initially differing significantly from that of a copolymer molecule formed towards completion of reaction.

This composition varies with degree of conversion. In extreme cases the monomer mixture is added to the reaction vessel in aliquots and the ratio of monomers to each other in each aliquot differs. However, for most purposes, little compensation is made for the changing composition. In many cases it is desirable to have a broad spectrum of copolymer composition (as well as molecular weight). To minimise extremes of composition it is possible to add all of a monomer to a reactor provided that it does not readily homopolymerise and then continuously add other monomers during the polymerisation reaction.

It is also possible to estimate the composition of the copolymer formed from a knowledge of the reactivity ratios and molar concentrations.

It was also shown in Chapter I that the composition of a copolymer at any instant could be related to reactivity ratios and concentrations of monomers by the equation:

$$\frac{d[M_1]}{d[M_2]} = \frac{[M_1]}{[M_2]} \cdot \frac{(r_1[M_1]+[M_2])}{(r_2[M_2]+[M_1])} = \frac{1+r_1}{1+r_2}$$

A few general examples may help to clarify the use of the equations involving reactivity ratios.

1. If one considers a copolymer of styrene and methyl methacrylate

$$r_1 = 0.50 \quad r_2 = 0.50 \quad r_1 r_2 = 0.25$$

This means that the polystyrene type radicals react with methyl methacrylate at almost twice the rate at which they react with styrene.

$r_2 = 0.50$ indicates that polymethyl methacrylate type radicals react with styrene at over twice the rate at which they react with methyl methacrylate, assuming both monomers are present at the same concentration.

2. If we consider a reaction mixture of styrene and butadiene

$$r_1 = 0.50, \quad r_2 = 1.4 \quad \text{and} \quad r_1 r_2 = 0.70$$

The ratio of the two monomers incorporated into the copolymer at the start of the reaction is

$$\frac{d[Styrene]}{d[Butadiene]} = \frac{1+0.50}{1+1.4} = 0.625$$

Thus rather more butadiene enters the copolymer than styrene, this means that the copolymer formed in the next instant of time has a different composition due to the changes in the ratios of styrene to butadiene from the initial one. This makes accurate predictions difficult, when one considers the many variables involved in polymer polymerisation.

The molar ratios initially present are altered and a different set of conditions need entering into the differential equation making computers ideally suited to calculating predicted compositions. Disproportionate consumption of one or more monomers alters monomer concentrations in the resulting monomer mixture. These concentrations are altering continuously from the commencement of polymerisation till conversion is in excess of 99%.

Another example is, 2-hydroxy propyl methacrylate copolymerised with styrene in the charged mole ratio of 70:30:

$$r_1 = 0.65 \qquad r_2 = 0.56 \qquad r_1r_2 = 0.364$$

$$\frac{d[M_1]}{d[M_2]} = \frac{[70]}{[30]} \times \frac{(0.65 \times 70) + 30}{70 + (0.56 \times 30)} \approx \frac{2}{1}$$

The initial copolymer formed would have a molecular ratio of 2:1 2−HPMA/styrene compared to an initial charge ratio of 2.33/1. As polymerisation proceeds the styrene is consumed at a slightly faster rate than 2−HPMA (r_1 and r_2 both < 1 and $r_1 > r_2$ i.e. $k_{1,2} > k_{1,1} : k_{2,1} > k_{2,2} : k_{1,2} > k_{2,1}$).

Polymer formed towards higher degrees of conversion will be 2−HPMA rich compared to that initially formed.

Continuous addition of monomers or variations in the rate of addition of different monomers would reduce this deviation from theoretical composition.

It should be noted that only two components (i.e. strictly copolymers) systems have been considered. Where terpolymers or high copolymers, which require six or more reactivity ratios are involved, the calculations become much more complex and can only be handled satisfactorily by a computer.

Termination

Chain termination of a growing polymer chain can only be accomplished, in the presence of a vinyl monomer, by removal of the free radical on the polymer chain. This can be brought about in several ways and the exact mechanism involved for a given polymer species depends to some extent on the chemical structure of the monomers and polymers involved.

a) Termination by combination is the simplest form of chain termination. This involves the free radicals of two propagating polymer chains coming together with mutual extinction of the radicals and the formation of a head to head linkage (see theory chapter, polymer tacticity), joining the two polymer chains together.

$$R - CH_2 - CHX - [CH_2 - CHX]_n - CH_2 - \overset{\bullet}{C}HX \ + \ XH\overset{\bullet}{C} - CH_2 - [XHC - CH_2]_m - XHC - CH_2 - R$$

$$\downarrow$$

$$R - CH_2CHX[CH_2 - CHX]_n - CH_2 - CHX - XCH - CH_2 - [XHC - CH_2]_m - XHC - CH_2 - R$$

Note: A head to head polymer link is formed.

b) Termination by disproportionation occurs when the free radical on the end of a propagating polymer chain abstracts a H atom from another propagating polymer chain causing mutual extinction of both free radicals. In this case the two chains do not combine but remain as separate entities, one chain containing an unsaturated end unit and the other a saturated end unit.

$$R - CH_2 - CHX[CH_2 - CHX]_n - CH - \overset{\bullet}{C}HX \ + \ XH\overset{\bullet}{C} - CH_2 - [XHC - CH_2]_m - XHC - CH_2 - R$$
$$\underset{H}{|} \qquad\qquad\qquad \downarrow$$

$$R - CH_2 - CHX - [CH_2 - CHX]_n - CH = CHX \ + \ XHC_2 - CH_2 - [XHC - CH_2]_m - XHC - CH_2 - R$$

Polymer with an unsaturated chain end   Polymer with saturated chain ends

A special case of inter molecular hydrogen abstraction can occur where the H atom is removed from the same chain as that of the abstracting free radical. This process is known as 'back biting' and is liable to occur where long chain polymers take up configurations which bring a labile H atom on the polymer chain into close proximity of the radical at the end of the chain.

Whether a polymer terminates by disproportionation or combination depends on the chemical configuration of the monomers involved.

Where there are no labile H atoms available then termination will be by combination, e.g. isobutyl styrene homopolymer. Where there are labile H atoms available termination occurs by both combination and disproportionation, e.g. acrylate monomers.

Where disproportion has occurred to a major extent in chain termination, the product will contain many polymer chains with unsaturated chain ends, or even with unsaturated sites along the polymer back bone as a result of intermolecular hydrogen abstraction. These double bonds are also capable of being reactivated by a free radical and hence propagation can continue in the presence of monomer.

c) Another termination mechanism involves removal of the free radical from the propagating chain by transferring it to another chemical species. With this type of reaction the free radical is not destroyed, just removed from the propagating chain.

There are various forms that this reaction can take. They are collectively referred to as 'transfer reactions' and are utilised in controlling the chain length (molecular weight) of a propagating species.

$$- [CH_2 - CHX]_n - CH_2 - \overset{\bullet}{C}HX + AB \qquad\qquad - [CH_2 - CHX]_n - CH_2 - CHXA + B^{\bullet}$$

Transfer reactions result in termination of a growing chain but not the extinction of the free radical.

Depending on the reactivity of the new radical so produced (B$^\bullet$ in the above equation) further reaction may take place using B$^\bullet$ as the initiating species.

AB can be a solvent, polymer or a modifier added as a chain transfer agent or an inhibitor, etc.

If B$^\bullet$ does initiate polymerisation then the overall polymerisation rate is unaffected by the transfer reaction, but of course the molecular weight is lower than that formed in a reaction without transfer.

The various types of transfer reaction are discussed in the following paragraphs.

Transfer to Solvent

Consider the polymerisation of styrene in carbon tetrachloride solvent.

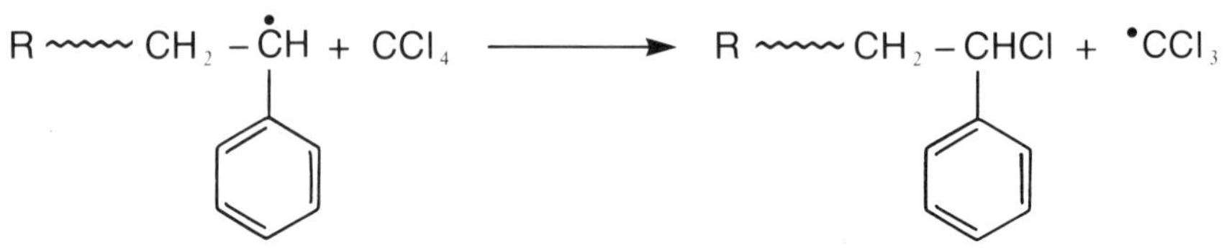

$$R \sim\!\sim\!\sim CH_2 - \overset{\bullet}{C}H + CCl_4 \longrightarrow R \sim\!\sim\!\sim CH_2 - CHCl + {}^\bullet CCl_3$$

| Polystyrene | Carbon tetrachloride | Radical |

Here the solvent has participated in the reaction resulting in the termination of the propagating chain and the formation of a new free radical species.

The $^\bullet CCl_3$ radical shown above is active and can reinitiate polymerisation.

$$^\bullet CCl_3 \quad + \quad CH \quad = \quad CH_2 \longrightarrow \overset{\bullet}{C}H - CH_2 - CCl_3$$

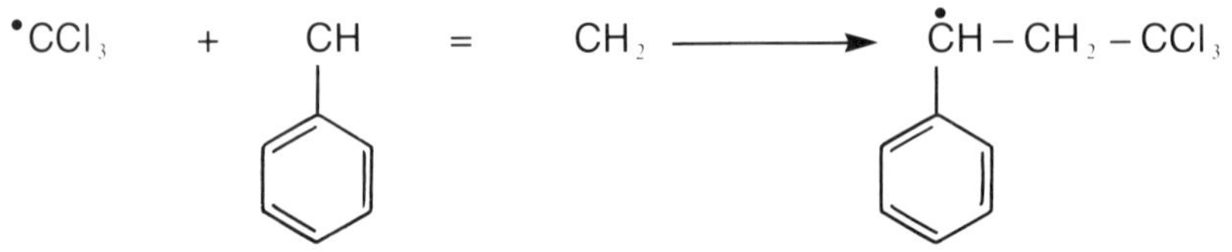

Thus the overall polymerisation rate is largely unaffected, but the premature termination of the polymer chain leads to a reduction in molecular weight of the polymer.

This type of reaction is very common and the ease with which a particular solvent can terminate a polymer chain and reinitiate another chain will depend on the chemical structure and hence, reactivity of the solvent, monomer and solvent radical. The ease of reaction is quantified by the transfer constant C_s which is the ratio of rate coefficient for transfer to the rate coefficient for propagation, i.e.

$$C_s = \left(\frac{k_s}{k_p} \right)$$

Where C_s = transfer coefficient

k_s = rate coefficient for transfer to solvent
k_p = rate coefficient for propagation of the solvent radical.

The C_s for a particular solvent varies with temperature and also solvent type.

Solvents are weak chain transfer agents and the table below lists some typical chain transfer constants; these are normally described by the term k_z which is defined as the ratio of the reactivity of a polymer radical towards the chain transfer agent to that of the monomer.

SOME VALUES OF CHAIN TRANSFER CONSTANTS AT 60°C

Monomer →	Acrylonitrile	M.M.A.	Styrene	Vinyl acetate
Solvent ↓				
Acetone	0.000095	0.00036	0.023	0.0016
Acetonitrile	0.00017	—	0.0023	0.0013
Aniline	1.05	0.0075	0.011	0.026
Benzene	0.00021	0.0014	0.00017	0.00016
Butanone (MEK)	0.00055	0.00089	0.028	0.0097
Carbon tetrachloride	0.000073	0.0043	0.57	1.18
Chloroform	0.00049	0.00089	0.00345	0.020
Ethyl acetate	0.00022	0.00027	0.0091	0.0004
Ethyl alcohol	—	0.00071	0.0085	0.0033
Tri-ethylamine	0.155	—	0.017	0.0490

In a free radical polymerisation where the rate of radical formation is kept constant, the degree of polymerisation is proportional to the concentration of monomer. Use of the term 'degree of polymerisation', which denotes the number of monomer units in the polymer, is convenient as it applies to all monomers. The degree of polymerisation multiplied by the molecular weight of the monomer gives the polymer molecular weight. Chain transfer to solvent not only reduces the molecular weight of the polymer formed initially but it also increases the polydispersity as polymerisation proceeds. The effect is illustrated in the figure below where the data has been calculated assuming that initial polymer would have a D.P. (degree of polymerisation) of 1000 if the solvent has no chain transfer activity.

EFFECT OF CONVERSION AND CHAIN TRANSFER ON
THE MOLECULAR WEIGHT OF SOLUTION POLYMERS

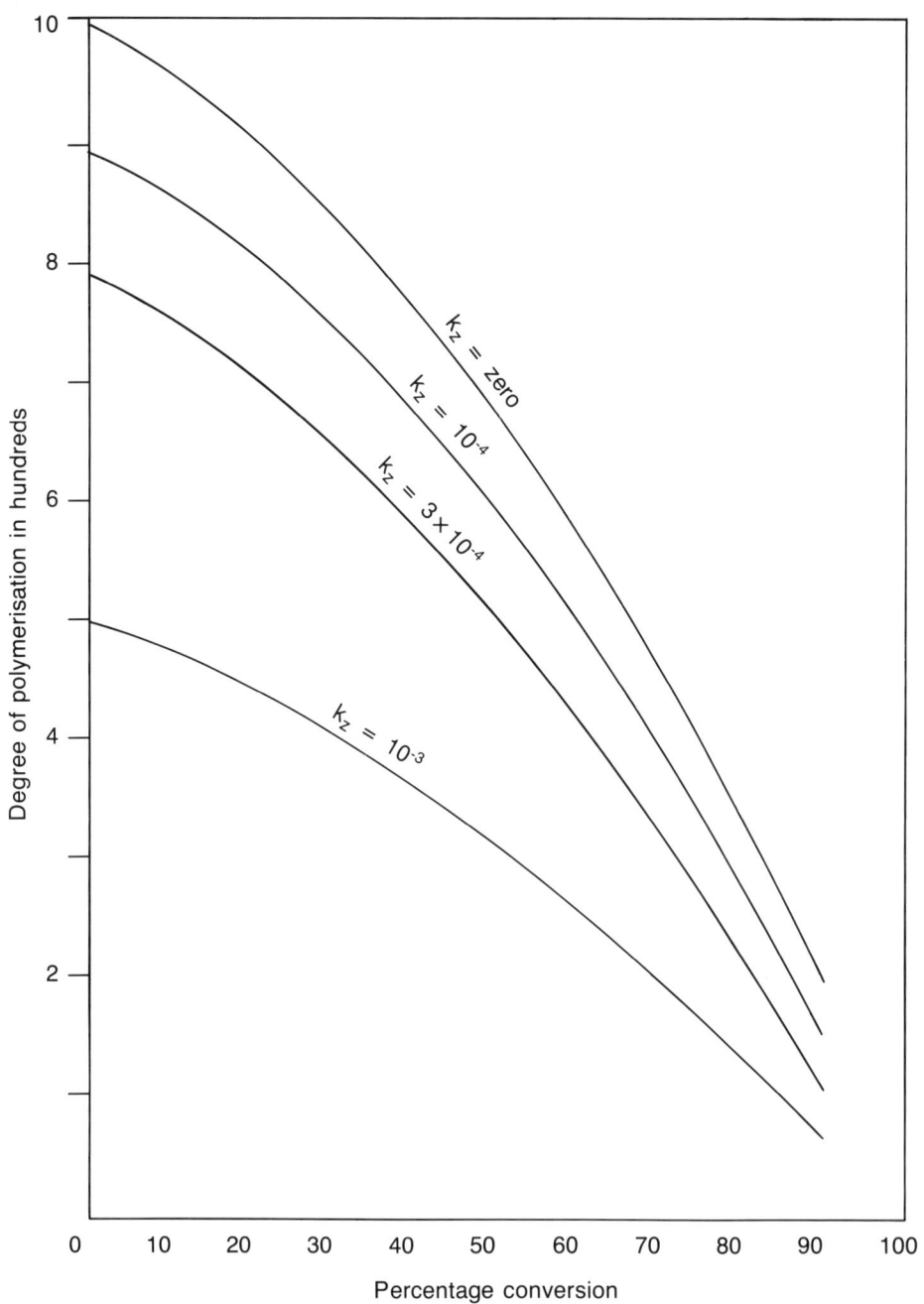

Degree of polymerisation in hundreds

k_z = zero

k_z = 10^{-4}

k_z = 3×10^{-4}

k_z = 10^{-3}

Percentage conversion

Unless a particularly low molecular weight is desired, solvents with low chain transfer activity are preferred; k_z values below 0.001 are normally used. The fall in monomer concentration as polymerisation proceeds also reduces the rate of polymerisation leading to longer reaction times. The use of as little solvent as possible allows both these effects of lower monomer concentration to be offset by the Tromsdorf effect (see later). Common industrial practice is to carry out the polymerisation at 60−70% solids when a near gel is produced which is only slowly turned by powerful stirring. The product is diluted to 50−60% solids to reduce the viscosity to a handleable level.

Transfer to Monomer

This involves H atom extraction and can occur in two ways. The first case is when transfer involves a labile H atom on the vinyl C, i.e.

$$R \sim\sim\sim CH_2 - \overset{\bullet}{C}HX + CH_2 = CHX \longrightarrow R \sim\sim\sim CH_2 - CH_2X + CH_2 = \overset{\bullet}{C}X$$

The initiated monomer radical can propagate a polymer chain, however, the chain formed will contain an unsaturated end group

$$CH_2 = \overset{\bullet}{C}X + nCH_2 = CHX \longrightarrow CH_2 = \overset{X}{\underset{|}{C}} - [CH_2 - CHX]_{n-1} - CH_2 - \overset{\bullet}{C}HX$$

Which can react further on reinitiation to give a branched chain polymer.

The second case is where the labile H atom is not on the vinyl C atom but in a side group, e.g. vinyl acetate polymer

Vinyl acetate radical   +   Vinyl acetate molecule

The free radical in the above example is on the side group not the vinyl C atom, and propagation again results in a polymer being produced with an unsaturated chain end.

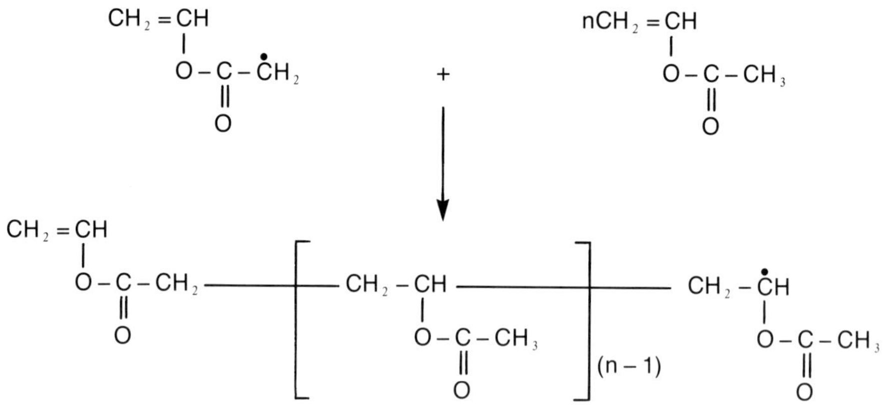

$$CH_2 = CH$$
$$|$$
$$O - C - \overset{\bullet}{C}H_2$$
$$\|$$
$$O$$

$$+$$

$$nCH_2 = CH$$
$$|$$
$$O - C - CH_3$$
$$\|$$
$$O$$

$$CH_2 = CH$$
$$|$$
$$O - C - CH_2 - \left[CH_2 - CH \right]_{(n-1)} - CH_2 - \overset{\bullet}{C}H$$
$$\| \qquad\qquad | \qquad\qquad\qquad\qquad |$$
$$O \qquad\qquad O - C - CH_3 \qquad\qquad O - C - CH_3$$
$$\qquad\qquad\qquad\quad \| \qquad\qquad\qquad\qquad \|$$
$$\qquad\qquad\qquad\quad O \qquad\qquad\qquad\qquad O$$

Unreacted double bonds in a polymer chain are a source of chemical weakness in a surface coating system particularly to attack by u.v. radiation when used in exterior coatings.

Transfer to Polymer

Again hydrogen atom abstraction is involved. The labile hydrogen may be on a neighbouring polymer chain or on the same polymer chain as the free radical, i.e. intermolecular or intramolecular hydrogen abstraction can occur.

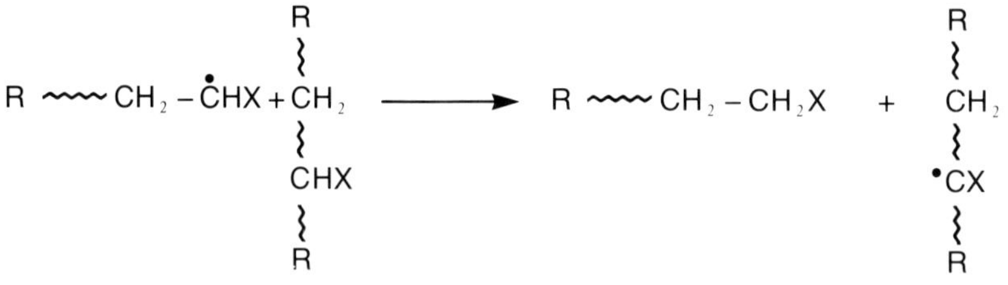

$$
R \sim\!\!\sim CH_2 - \overset{\bullet}{C}HX + CH_2 \quad\longrightarrow\quad R \sim\!\!\sim CH_2 - CH_2X \quad + \quad CH_2
$$

with side groups:

$$
\begin{array}{ccc}
R & & R \\
\} & & \} \\
CH_2 & & CH_2 \\
\} & & \} \\
CHX & & {}^\bullet CX \\
\} & & \} \\
R & & R
\end{array}
$$

The re-initiated polymer can proceed to incorporate more monomer units to produce a polymer unit with a lengthy side chain or branch.

$$
\begin{array}{ccc}
\text{R} & & \text{R} \\
\{ & & \{ \\
\text{CH}_2 & + \quad n(\text{CH}_2 = \text{CHX}) \longrightarrow & \text{CH}_2 \\
\{ & & \{ \\
{}^\bullet\text{CX} & & \text{CX} - [\text{CH}_2 - \text{CHX}]_{(n-1)} - \text{CH}_2 - \overset{\bullet}{\text{C}}\text{HX} \\
\{ & & \{ \\
\text{R} & & \text{R}
\end{array}
$$

The re-initiation of terminated polymer chains can lead to the formation of branched polymers.

The frequency with which termination of transfer to polymer occurs can be measured by determining the amount of branching that occurs per unit length of polymer chain. Monomers with relatively short side chains such as styrene, methyl methacrylate, etc., show little tendency to form branches, whilst monomers such as lauryl methacrylate or vinyl acetate have very high branching coefficients and branching occurs easily during polymerisation.

Transfer to Modifier

This type of reaction is similar to the transfer to solvent mechanism. Modifiers or chain transfer agents are species that contain a labile atom (usually a hydrogen or halogen atom) which is abstracted by the growing polymer chain. In this reaction, propagation of the chain is stopped and the free radical transferred to the modifier. The modifier radical so formed can then initiate propagation of a new polymer chain. Because its overall effect is to transfer the free radical from one polymer chain to another the modifier is known as a chain transfer agent. Examples of typical chain transfer agents are: carbon tetra bromide, t-butyl mercaptan, and ethanethiol.

The effectiveness of a chain transfer agent is measured by how easily it terminates a growing polymer chain and then re-initiates another chain, and this will vary with monomer type. Generally the most effective chain transfer agents are long chain alkyl mercaptans and these are extensively used to control the molecular weight of vinyl polymers during manufacture.

$$
\text{R} \sim\!\sim\!\sim \text{CH}_2 - \overset{\bullet}{\text{C}}\text{HX} \quad + \quad \underset{\text{mercaptan}}{\text{RSH}} \longrightarrow \text{R} \sim\!\sim\!\sim \text{CH}_2 - \text{CHX} + \text{RS}^\bullet
$$

$$
\underset{\text{Mercapto radical}}{\overset{\bullet}{\text{R}}\text{S}} \quad + \quad \text{CH}_2 = \text{CHX} \longrightarrow \text{RS} - \text{CH}_2 - \overset{\bullet}{\text{C}}\text{HX}
$$

An approximation of the molecular weight produced in the presence of a chain transfer agent can be obtained from Mayo's equation:

$$
\frac{1}{P} = \frac{C_s [S]}{[M]} + \frac{1}{P_o}
$$

Where P = degree of polymerisation (average number of monomer units per chain), in presence of the C.T.A. (Chain Transfer Agent).

P_0 = degree of polymerisation without C.T.A.
C_s = chain transfer constant for the modifier concerned
[S] = molar concentration of modifier present
[M] = molar concentration of monomer present

The effectiveness of a C.T.A. to transfer a free radical from a growing chain is measured by the transfer constant C_s.

Comparison of C_s for various monomers of some of the more common C.T.A.'s is shown below.

| CHAIN TRANSFER AGENT (C.T.A.) | HOMOPOLYMER AT 60°C | | |
	Methyl methacrylate	Styrene	Methyl acrylate
Carbon tetrabromide	0.27	2.2	0.41
Butanethiol	0.66	22	1.69
t-butyl mercaptan	0.18	3.6	—
Ethyl mercaptoacetate	0.63	58	—

The C_s of a C.T.A. varies with reaction temperature and the monomer type involved.

CONVERSION OF MONOMER TO POLYMER

It is essential to remove the large quantities of heat that are liberated during polymerisation. It is therefore imperative that reaction conditions should favour a uniform conversion of monomer to polymer in order that the heat of reaction can be dissipated in a controlled manner.

The majority of commercial polymerisations employ a continuous addition technique. This technique requires monomer and initiator to be added to the reactor at a rate approximately equal to the rate of conversion to polymer. In this way a build-up of unreacted monomer is prevented.

The quantity of unreacted monomer present at any instant must be carefully monitored and corrective action taken immediately it deviates from an acceptable level. The unreacted monomer present during the polymerisation can be quantified by measurement of the non-volatile content and comparison of this result with the calculated theoretical non-volatile content. Examples of theoretical and typical practically obtained curves are shown below.

ALL MONOMER AND INITIATOR INITIALLY PRESENT ('All In')

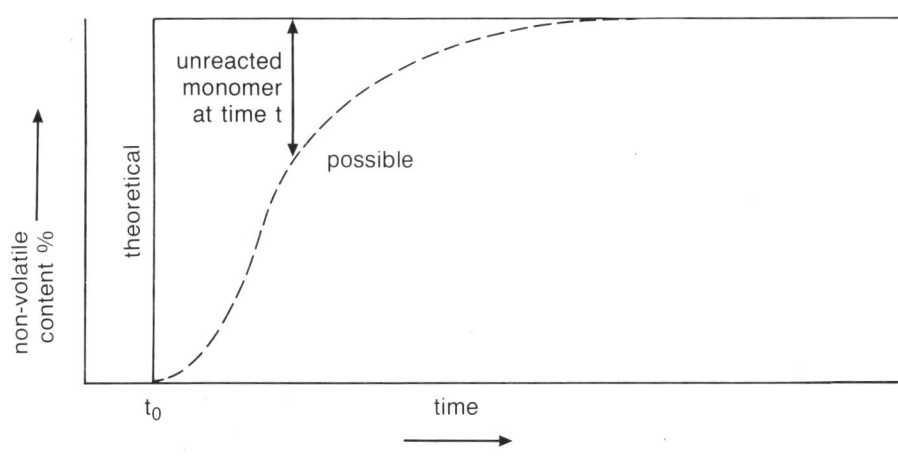

For an 'all in' process the possible curve is about as good as one could realistically expect. Therefore, it is no surprise that the initial charge is reduced to minimise uncharged monomer and when that has reacted, further monomer is added.

CONTINUOUS ADDITION (1)

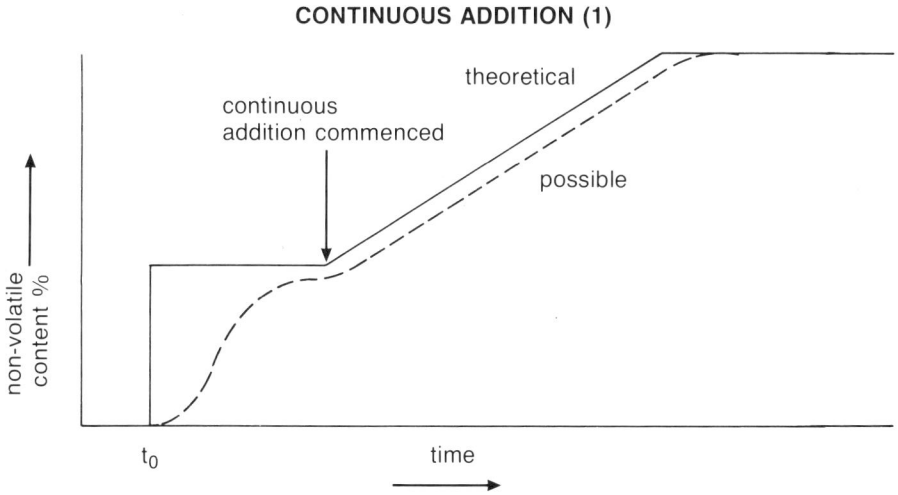

An initial aliquot of monomer (perhaps 25–30% of the total charge) is added. When this has been converted, continuous addition of the remainder commences.

If this is uniform then the theoretical conversion will follow the above graph.

CONTINUOUS ADDITION (2)

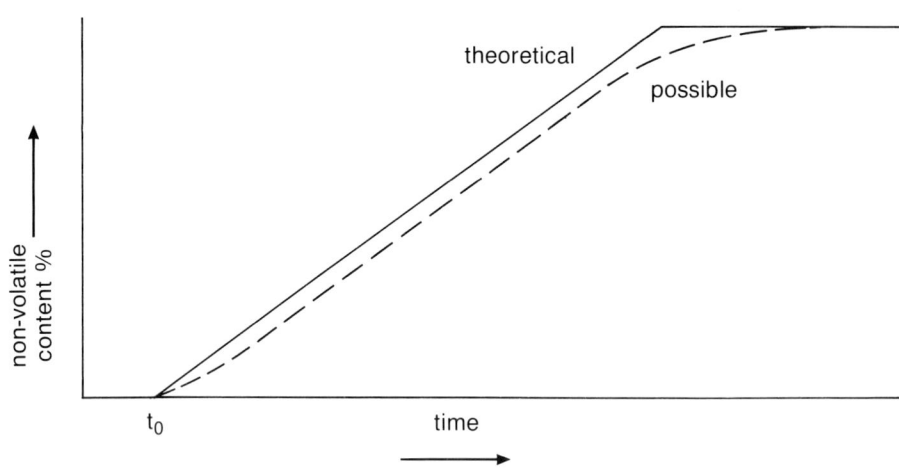

This diagram represents the theoretical n.v.c. for a continuous uniform monomer addition without any aliquot of monomer being initially charged to the reactor.

All of the above cases, are under normal conditions, generally acceptable. However, we will now examine examples of possible unacceptable curves.

'ALL IN' ADDITION

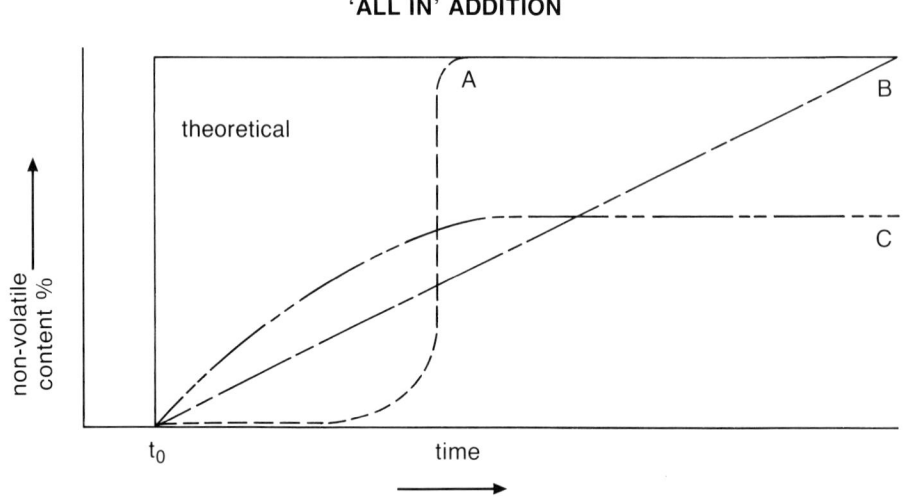

Curve A although reaching theoretical n.v.c., indicates all the monomer polymerises in a very short period of time. For large scale polymerisation this would probably be unacceptable because the heat generated would be more than the reactor could remove and the resulting rapid increase in viscosity accompanying the conversion of monomer to polymer would make the heat transfer even worse. The only common exception might be suspension polymerisation, where the heat sink (water) is normally double the weight of monomer present.

Curve B allows an unacceptable amount of unreacted monomer to remain during the early and middle stages of polymerisation. This system would be ideally suited to a continuous addition technique.

Whilst Curve C appears initially promising, it shows that some of the monomers are not going to polymerise and the system should not be scaled up. Reformulation and process modifications are required.

For continuous additions, examples of unacceptable curves might be:

CONTINUOUS ADDITION (1)

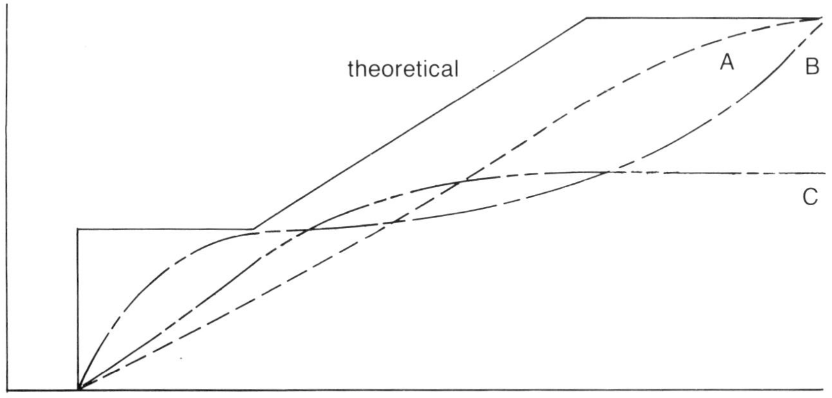

The process of Curve A is better suited to a slower continuous addition rate, possibly without (or with a smaller) initial monomer aliquot or further delayed commencement of the continuous monomer addition. Comparison of the continuous theoretical and possible curve might be acceptable if the continuous addition were delayed.

Whilst Curve B demonstrates that theoretical conversion has occurred prior to commencement of the continuous addition, the rate of the continuous addition is disproportionately faster than the polymerisation rate. This process would probably be acceptable if the addition rate were decreased. It may be possible to reduce the difference by increasing the reaction temperature or adding further initiator or different initiators.

Similar comments apply to Curve C as for Curve C previously.

To determine the potential heat generated due to instantaneous polymerisation of all the unreacted monomer present, the following approach can be used:

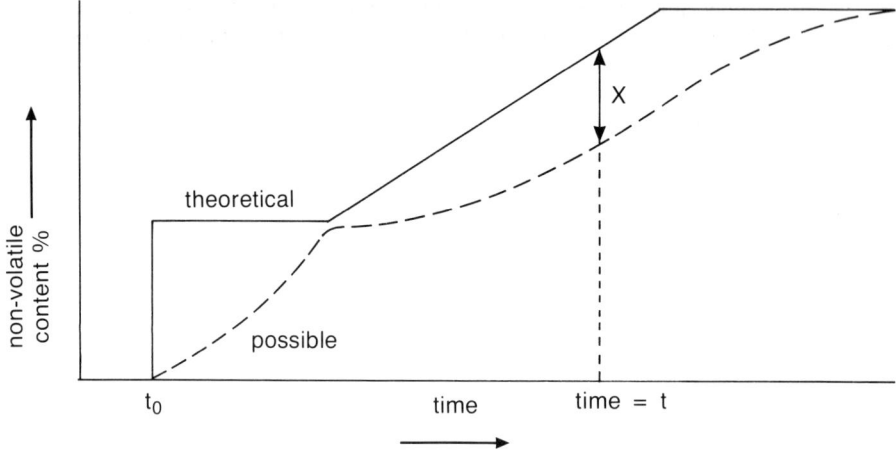

X represents the % of unreacted monomer. The total monomer charged to time t is known. Therefore the weight of unreacted monomer is calculated by:

$$\frac{\text{Theoretical \% } - \text{ Actual \%}}{100} \times \text{(total weight of monomer charged to time t)}$$

This weight can then be converted to heat of reaction using the method outlined earlier. The thermal capacity of the reactor and partially charged (and reacted) components must be estimated. A comparison with the rate of cooling of the reactor will readily show if there is likely to be an uncontrollable exothermic reaction. Allowance must be made for poor thermal transfer due to viscosity increases, organic liquids and inefficient agitation.

This approach can also be used to reduce times of reaction by increasing the rate of continuous addition until the residual monomer is unacceptably high.

As general rules, do not scale up if there is any doubt about the level of unreacted monomer and the cooling capabilities of the reactor, and exercise extreme care in reducing addition times or increasing the initial monomer aliquot.

POLYMER PROPERTIES

Mechanical, physical and chemical properties of polymers relevant to their end use will be discussed under applicational headings. The two general properties of polymers to be discussed here are molecular weight and glass transition temperatures.

Molecular Weight

The number and weight average molecular weights have been mentioned in Chapter I, Volume I. In calculations, the degree of polymerisation (D.P.) can be derived from a knowledge of the relevant reaction constants. The D.P. is a number average value which is derived from experimental molecular weight averages. Any method for determining molecular weights needs calibration and this involves two techniques. The first uses a living radical, often sodium naphthalene or butyl lithium, to produce a polymer of known $\bar{M}_n$ with a very narrow range of molecular weights. Several polymer standards of this type can be purchased from the National Physical Laboratory. The other technique is the fractionation of a polymer into narrow molecular weight ranges by fractional precipitation. A polymer solution in a good solvent is treated with a non-solvent until a faint opalescence occurs. Addition of a small amount of non-solvent causes precipitation of the highest molecular weight polymer. Subsequent small additions of non-solvent precipitate successively lower molecular weight fractions.

Early calibrations were based on fractionated samples whose molecular weight was determined by osmometry which can measure molecular weights up to 100,000. Osmometry gives $\bar{M}_n$ directly. The light scattering method is widely used in universities and can measure $\bar{M}_w$ for molecular weights above 30,000. The ultra centrifuge method measures the z average and is most useful in the molecular weight range 100,000 to 500,000. The z average is defined as:

$$\bar{M}_z = \frac{\Sigma (n_i \, M_i^{\,3})}{\Sigma (n_i \, M_i^{\,2})}$$

Commercial laboratories generally use a viscometric method to measure molecular weight. This gives an average between $\bar{M}_n$ and $\bar{M}_w$.

The method involves measuring the viscosity of solutions of different concentrations in a U-tube viscometer as well as the viscosity of the solvent. The use of a thermostatically controlled water bath to give good temperature control is essential for accurate work.

The theory this method is based upon can be briefly outlined as follows:

The increase in viscosity from that of the solvent (ηo) to that of the solution (η) due to polymer can be defined as the specific viscosity (ηsp).

$$\eta sp = (\eta - \eta o)/\eta o$$

The reduced viscosity (ηr) depends upon polymer concentration (c), viz:

$$\eta r = \eta sp/C$$

At infinite dilution (i.e. zero concentration) the intrinsic viscosity [η] is obtained, by plotting ηrel against C and entrapolating to zero concentration.

The intrinsic viscosity is the limiting value of the reduced viscosity at zero concentration, as shown in the figure on the next page.

[η] can also be obtained by plotting either
ln ($\eta sp/c$)   or   ln ($\eta/\eta o)/c$ against C

There is an empirical relationship relating [η] to molecular weight.

$$[\eta] = kM^a$$

where **k** and **a** are constants for a given polymer solvent system and can be obtained for many homopolymers from reference books like the Polymer Handbook.

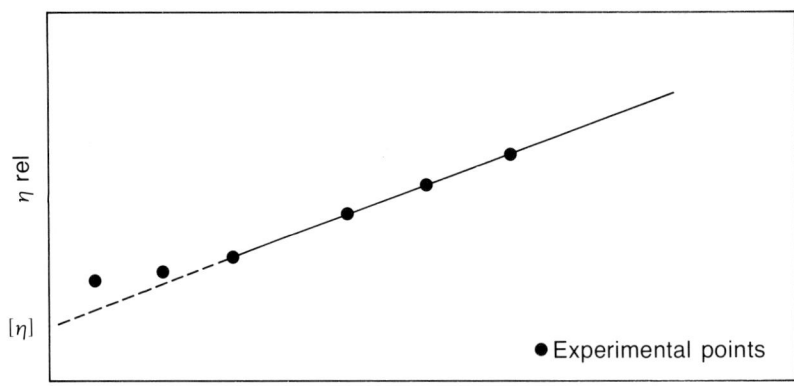

Concentration (grams polymer per decilitre)

A typical intrinsic viscosity plot

For flexible polymer chains the constant **a** is normally in the range 0.5 to 0.8, whereas for stiffer chains the constant **a** is 0.8 to 1.0. As a generalisation **a** can be considered to be between 0.5 to 1.5, and is often referred to as the Mark Houwink constant.

The determination of accurate intrinsic viscosities is tedious, time consuming, and requires an extremely meticulous experimental approach. Very fine capillary tube viscometers are used and they can easily be blocked by dust or other extraneous matter. Careful filtration through sintered glass is essential. For the highest quality work it is important that solution concentrations are not altered inadvertently during filtration, dilution or measurements. Blowing (pressure) rather than sucking (vacuum) techniques should be used to avoid evaporation of solvent when for example raising the solution in the side arm prior to timing. Accurate temperature measurement and control are essential.

In quality control work it is normal to determine the solution viscosity at one concentration only, generally $0.5-1.0$ g per $100 \, cm^3$, and to calculate the intrinsic viscosity from a knowledge of the slope of the calibration curve.

In recent years gel permeation chromatography has been used increasingly in industrial laboratories. A polymer solution is pumped through a column filled with a gel which contains holes of a known size. Large molecules cannot enter the holes and therefore pass straight along the column with no delay. The smaller the molecule the more likely it is to enter a hole and each time it does so its passage along the column is interrupted. The smaller the molecular weight the more slowly the compound travels along the column. This method is quick and reproducible. It also gives both the molecular weight and its range. The g.p.c. method gives a weight average molecular weight, but it is difficult to obtain an absolute figure because it relies on the measurement of hydrodynamic volume. Polymer dissolved in a poor solvent tends to coil into a tight sphere, while a polymer in a good solvent has more spread out polymer molecules. Primary standards are available for calibration of many homopolymers for which accurate calibration is possible in any solvent. Primary calibration is not possible for copolymers and the g.p.c. method can only give relative values. However, the method is still valuable as a quality control tool for reproducibility from batch to batch.

Glass Transition Temperature

Melting, boiling and sublimation are the three changes of state which are known as first order transitions. The melting temperature T_m is a solid/liquid transition. Some polymers degrade at temperatures below T_m. Second order transitions generally involve changes in the solid state. The energy changes involved are smaller than the latent heat's of first order transitions. In polymer physics the glass transition temperature is the most important of the second order changes and it represents the change from glass to rubber; above the Tg a polymer is flexible and has a high elongation at break; whilst the glass form which exists below the Tg is brittle. The glass transition temperatures (Tg) for some homopolymers are shown in the table below.

Tg VALUES FOR SOME COMMON HOMOPOLYMERS	
Polymer	Tg °K
Polyethylene	148
Poly(cis-butadiene)	171
Poly(transbutadiene)	215
Poly(acrylic acid)	379
Polyacrylamide	438
Poly(ethyl acrylate)	249
Poly(butyl acrylate)	219
Poly(2-ethylhexyl acrylate)	223
Poly(methyl methacrylate)	378
Polymethylene	155
Polypropylene	260
Polystyrene	373
Poly(vinyl acetate)	305
Poly(vinyl chloride)	354
Poly(vinylidine chloride)	255
Poly(acrylonitrile)	383
Poly(n-butyl methacrylate)	295
Poly(2-ethylhexyl acrylate)	203
Poly(2-hydroxy propyl acrylate)	266
Poly(ethyl methacrylate)	338
Poly(methyl acrylate)	282
Poly(2-hydroxy ethyl acrylate)	258
Poly(2-hydroxyethyl methacrylate)	328
Poly(vinyl pyrrolidone)	327
Poly(iso-butyl methacrylate)	321
Poly(2-hydroxy propyl methacrylate)	346

Almost any property or characteristic which changes sufficiently with temperatures may be used to determine the glass transition temperature. Properties such as specific volume, refractive index, specific heat, thermal distortion, flexibility, rigidity, viscosity, di-electric properties, etc., can be measured. Differential Scanning Calorimetery (D.S.C.) is among the more modern methods for quick accurate determinations.

It is possible to predict approximate Tg's of copolymers from a knowledge of Tg's of the homopolymers formed from the constituent monomers. Either mole or weight fractions can be used in the calculations. Different workers use different quantities.

The mole fraction approach is summarised as follows; the Tg of a copolymer of known composition can be calculated from the Tg values of its component monomers:

$$\frac{1}{Tg} = \frac{X_1}{Tg_1} + \frac{X_2}{Tg_2} + \frac{X_i}{Tg_i} \quad \dots\dots\dots\dots\dots\dots$$

where X_1, X_2 and X_i are mole fractions of components 1, 2 and i, and Tg_1, Tg_2 and Tg_i are Tg's (°Absolute) of components 1, 2 and i.

The copolymer described in the emulsion examples can be used to illustrate the method of calculation.

Firstly, the mole fractions must be found.

Moles of vinyl acetate = 400/86 = 4.651 moles.
(molecular weight of VA = 86).

Moles of 2–EHA = 100/128 = 0.781 moles.
(molecular weight of 2–EHA = 128).

Mole fraction of vinyl acetate $= \dfrac{4.651}{4.651 + 0.781} = 0.856$

$\dfrac{1}{Tg} = \dfrac{0.856}{305} + \dfrac{0.144}{223} = 0.00281 + 0.00065 = 0.00345$

Tg = 290°K or 17°C

If the calculation had been worked out using weight fractions, the answer would be 11°C which represents a difference of 6°C.

The weight fraction approach is summarised as follows: approximate values for the Tg's of copolymers can be calculated from the Tg's of the homopolymers used.

$$\frac{1}{Tg} = \frac{WM_1}{Tg_1} + \frac{WM_2}{Tg_2} + \frac{WM_i}{Tg_i}$$

Where Tg is °K, WM_x is the weight fraction of each monomer and Tg_x is the glass transition temperature of the respective homopolymer: 70% Styrene (Tg = 100°C); 30% Butyl acrylate (Tg = −56°C).

$$\frac{1}{Tg} = \frac{0.7}{(273+100)} + \frac{0.3}{(273-56)}$$

$$\frac{1}{Tg} = 0.001877 + 0.001384 = 0.003259$$

$$Tg = 307°K \quad or \quad 34°C$$

This example illustrates the effect on Tg of blending a soft monomer (Tg below room temperature) with a hard monomer (Tg above room temperature).

The above equation assumes a linear relationship between weight composition and the Tg, and is valid in many cases.

However, some combinations, including vinylidene chloride, do show differences between the calculated and theoretical values.

As the equations for calculating Tg are approximations, the differences introduced by using mole or weight fractions are under normal circumstances negligible in comparison to other differences between actual and calculated Tg's inherent in the assumptions used. It should be remembered that these equations are only used as guidelines for the initial formulations. Weight fractions are more convenient to use and as such the majority of resin formulators use them rather than mole fractions. Having made initial formulations it is normally necessary to modify the type or ratios of hard to soft comonomers to obtain other performance properties.

These equations are meaningful if they are used to give relative comparisons of Tg's from a different selection of comonomers. The initial types and weights of comonomers can be deduced. The next step depends upon the experience and judgement of the resin chemist, combined with the constraints supplied by the end user. It should be possible to reduce the starting formulations to a maximum of two or three combinations of monomers.

It has been suggested that the Fox Johnston equation gives a reasonable correlation between theoretical and practical values.

For a Terpolymer

$$\frac{1}{Tg} = \frac{W_{AA}}{Tg_{AA}} + \frac{W_{BB}}{Tg_{BB}} + \frac{W_{CC}}{Tg_{CC}} + \frac{W_{AB}}{Tg_{AB}} + \frac{W_{AC}}{Tg_{AC}} + \frac{W_{BC}}{Tg_{BC}}$$

W_{AA}, W_{BB}, W_{CC} = Weight fraction of AA, BB or CC, monomers in polymer

W_{AB}, W_{AC}, W_{BC} = Weight fraction of AB, AC or BC diads in polymer

$Tg_{AA}, Tg_{BB}, Tg_{CC}$ = Glass transition temperatures for homo-polymers A, B or C.

Tg_{AB}, Tg_{AC}, Tg_{BC} = Glass transition temperatures for alternating copolymers AB, AC or BC.

In the equation diads AB = BA, AC = CA and BC = CB.

At the Tg of a polymer significant rotation in the backbone or side chains occurs. This is due to thermal excitement. Polymer properties alter considerably as it passes through this temperature zone. The polymer is rigid, glassy, hard and brittle at temperatures lower than its Tg. While at temperatures higher than its Tg it becomes softer, rubbery and flexible.

A polymer which has a Tg below 25°C tends to be soft and flexible at ambient temperatures, with the possibilities of cold flow occurring particularly under warm conditions. On the other hand a polymer with Tg above 25°C is hard and brittle. The best compromise for toughness is often achieved using polymers whose Tg is near to 25°C.

Cross linking has little effect on hardness at temperatures below the Tg. However, at temperatures which exceed the Tg cross linking greatly increases the polymer hardness. Side chains affect the Tg of the resultant polymer. In general the longer the side chain, the lower the Tg.

Extra CH_2 groups in substances such as the methacrylates give additional steric hindrance, which raises the Tg.

Additionally, a branched group on the side chain raises the Tg of the polymer. Tg increases slightly with increasing molecular weight.

The following diagrams illustrate some of the relationships between Tg and mechanical properties.

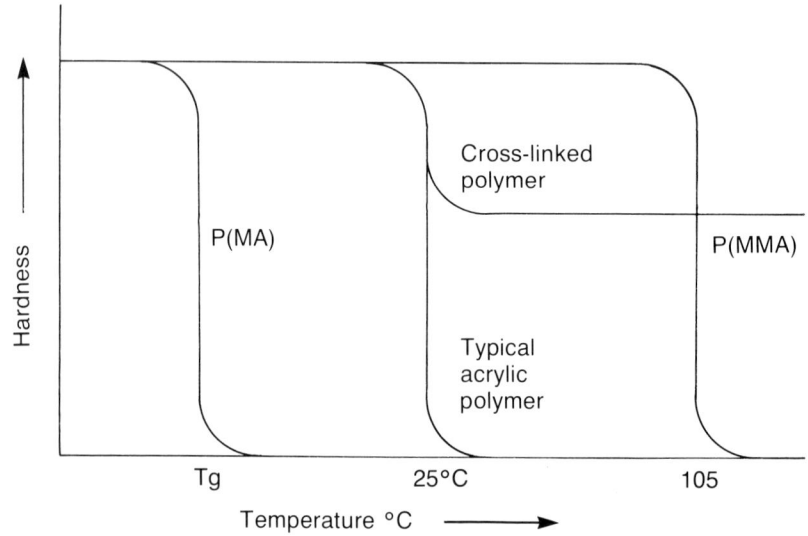

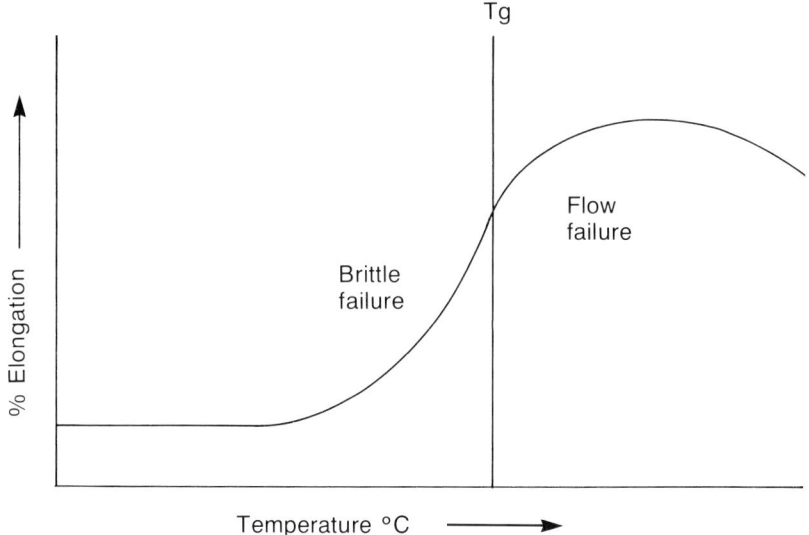

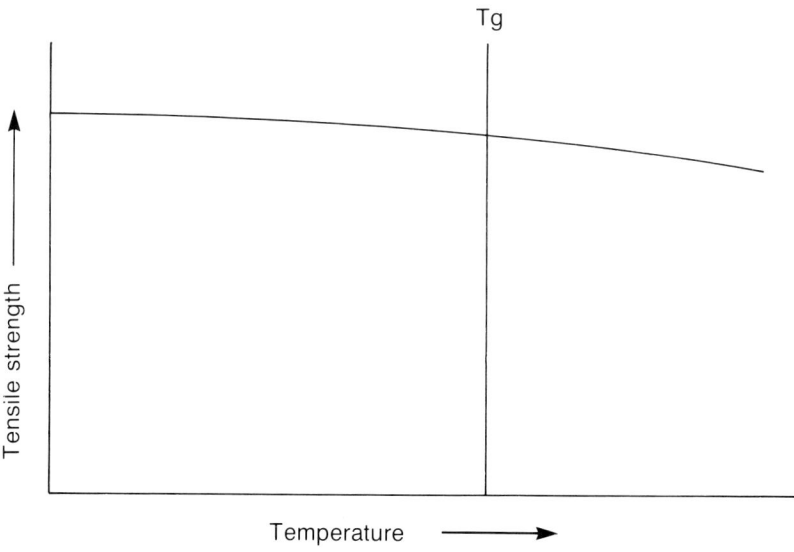

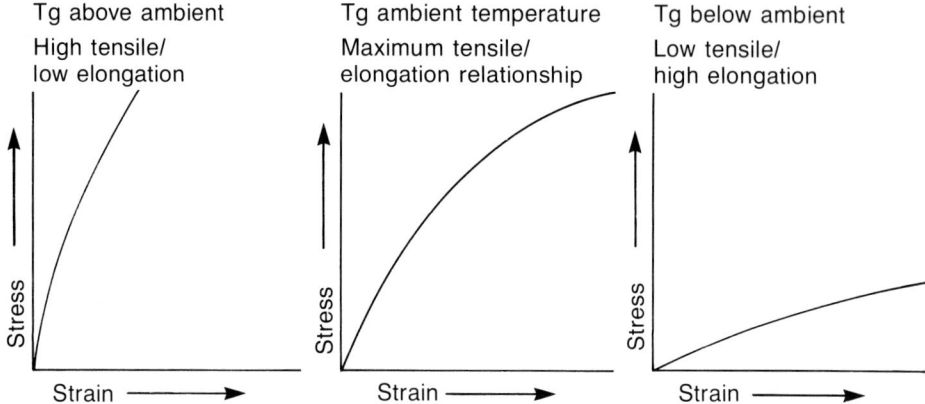

Literature quoted values illustrate changes in strength and elongation with decreasing Tg.

Polymer	Tensile strength (PSI)	Elongation (%)	Tg
Poly(methyl methacrylate)	9000	4	+ 105°C
Poly(methyl methacrylate)	5000	7	+ 65°C
Poly(butyl methacrylate)	1000	250	+ 22°C
Poly(methyl acrylate)	1000	750	− 9°C
Poly(methyl acrylate)	33	1800	− 22°C
Poly(butyl acrylate)	3	2000	− 56°C

TECHNIQUES OF POLYMERISATION

Addition polymerisation is brought about commercially by one of five techniques, although only two, the solution and the emulsion polymerisation techniques are of major importance in producing resins for surface coating applications.

Bulk Polymerisation

This is the simplest of the available techniques. In bulk polymerisation monomer and initiator or catalyst are mixed together and heated to reaction temperature.

The initiators used may be free radical, cationic or anionic types.

This technique produces very high molecular weight polymers and is not normally used to produce resins for use in surface coating applications. It is normally restricted to the commercial preparation of the so called 'high polymers' destined for use in moulding and forming applications. Polyethylene, polypropylene, polystyrene, polymethyl methacrylate, are all produced commercially by variations of the bulk polymerisation technique. The reaction rate is difficult to control due to the heat of polymerisation and auto-acceleration. High viscosities, and low thermal conductivity make such conversions of all monomers difficult, and hence give rise to a free monomer content which is also difficult to remove.

Conversion of methyl methacrylate is accompanied by a 21% shrinkage factor on polymerisation which is difficult to accommodate. The polymerised acrylic may be cast into sheets, extruded or ground into suitable sized particles.

ADVANTAGES
  i) Polymers made using the bulk polymerisation technique are very 'pure' since only catalyst and monomer are normally present in the reaction mixture.
  ii) High molecular weight polymers can be formed due to the absence of chain terminating species in the reactant mixture.

DISADVANTAGES
  i)   High temperatures are encountered since there is little facility for removal of the exothermic heat of reaction.
  ii) As the polymerisation proceeds the viscosity of the polymer in solution in the unreacted monomer rises rapidly and adequate agitation is impossible. This increase in viscosity effects the propagation and termination rates. As a result polymers with a very wide molecular weight distribution are produced.
  iii) Occasionally a particular polymer may be insoluble in the unreacted monomer and is thus precipitated from solution as soon as it is formed.

Because of the above disadvantages, bulk polymerisation is often carried out as a continuous or semi-continuous process. Polymerisation is normally only carried out to relatively low degrees of conversion, and the unreacted monomer recycled. Bulk polymerisation finds little application for surface coating resin preparations.

Solution Polymerisation

This technique involves the polymerisation of vinyl monomer in the presence of an organic solvent. In order for this technique to be effective both monomer and polymer must be soluble in the chosen solvent. A mixture of monomer, initiator, solvent (and chain transfer agent when low molecular weight is required) are heated together to reaction temperature, which is usually the reflux temperature of the solvent, or solvent mixture. Polymerisation of the monomer takes place to give a solution of polymer in the solvent. The polymer solution can be used directly in surface coating systems either on its own, thermoplastic types, or in conjunction with other resins as co-cure systems, thermosetting types.

MOLECULAR WEIGHT

Polymers made by the solution polymerisation process are relatively lower in molecular weight than other polymer types. The viscosity of the polymer solution is related to the solubility characteristics of the solvent and polymer, and also to the molecular weight of the polymer (see viscosity average molecular weight section in the theory chapter).

The molecular weight of the polymer is determined by the transfer constants of the solvent and chain transfer agent present and also by the half life ($t_{\frac{1}{2}}$) of the initiator used and the reaction temperature employed.

As reaction temperature is increased the number of free radicals generated by thermal decomposition of the peroxide increases. This results in the initiation of more polymer chains containing few monomer units. Thus the molecular weight and hence the solution viscosity of the polymer is decreased. As the reaction temperature is decreased the reverse occurs. Thus by careful selection of initiator type and reaction temperature a degree of control can be exercised over the polymer viscosity. However, the changes in molecular weight produced by this method are small compared to the gross changes that can be effected by use of an efficient chain transfer agent such as t-dodecyl mercaptan.

The molecular weight distribution is effected by the precise method employed to prepare the polymer. Two distinct techniques are available to bring about polymerisation and each has an effect on the molecular weight distribution and on the viscosity of the final solution polymer.

ONE SHOT PROCESS ('All In')

This involves charging all the reactants together with solvent, initiator and modifier into the reactor and heating to the optimum breakdown temperature of the initiator employed. As propagation occurs the temperature of the reactants is allowed to rise to the reflux temperature of the solvent present.

The reactants are then held at this temperature under reflux using the latent heat of vaporisation of the solvent (also the monomer) and reactor water cooling to remove excess heat of reaction.

The course of the reaction is monitored by viscosity and non-volatile content measurements, and when conversion of the monomer is complete the reactor is cooled. A wide molecular weight distribution is obtained using this technique. Since the concentration of initiator and modifier fall as the reaction proceeds, polymer chains produced later in the process are not subjected to the same frequency of termination reactions and are thus much higher in molecular weight than polymer produced earlier in the process.

DRIP-FEED PROCESS (Continuous Monomer Addition)

This involves 'drip-feeding', a premixed charge of monomer, modifier and initiator into solvent which has been preheated to reaction temperature. Ideally as the premix droplet comes in contact with the hot solvent the initiator decomposes to free radicals which immediately initiate propagation of the monomer. The polymer formed is dissolved in the solvent and is dispersed. In practice reaction does not occur immediately the monomer contacts the hot surface and there is a slight build up initially of unreacted monomer. However, this can be maintained at a relatively low concentration by correct choice of reaction conditions. The addition of a small amount of extra initiator after the premix addition is completed, is sufficient, normally, to complete the conversion of all the monomer present.

At any time during the polymerisation the relative amounts of monomer, initiator, modifier and solvent are all reasonably constant. As a result polymers with a narrow molecular weight distribution are formed.

The premix feed rates are a critical factor in producing poly-dispersed material without a build up of unreacted monomers which could lead to an uncontrollable exotherm and certainly to a wide spread of molecular weight. Therefore, feed rates have to be carefully determined during laboratory process development. Commercial feed rates vary between two and ten hours, depending on monomer type and reaction conditions. Where highly reactive monomers or initiators with very short $t\frac{1}{2}$ are employed it is usual to keep the initiator separate from the monomer premix and add it separately at a predetermined rate throughout the monomer addition. This prevents the possibility of initiation occurring in the premix tank.

The use of the drip feed process allows more latitude in the preparation of constant composition copolymers. By constantly varying the composition of the monomer feed during addition, the differences in reactivity between monomer types can be partly countered to produce copolymers approaching constant composition.

In practice, for surface coating polymers, constant copolymer composition is not required. Since most copolymers are manufactured by the batch process the overall composition of the final solution is constant and it is of little consequence that individual polymer chains may not have exactly the same monomer unit composition.

The differences between a wide and narrow molecular weight distribution (i.e. polydispersed and monodispersed) polymer solutions is generally evident in their rheology and in the film performance characteristics. In general, polydispersed systems are more flexible and have lower solution viscosities. However, since these properties are also influenced by other factors such as copolymer composition, solvent type and overall molecular weight, solution polymers made by either process are acceptable, although the drip-feed process allows a stricter control on process conditions to be exercised and this is often desirable when dealing with highly exothermic chemical reactions and when product consistency is required.

MONOMERS

Any vinyl monomer that forms a homopolymer or copolymer which is soluble in the solvent used may be polymerised by this process. Basically, two types of polymer are produced for surface coating systems by the solution polymerisation process.

These are thermoplastic types and thermosetting types. They are distinguished by the monomers utilised.

Simply, thermosetting types contain monomers with functional pendent groups which can be cured with themselves or with functional groups on other polymer types. Typical of the functional monomer types used commercially are acrylamide and hydroxy ethyl acrylate. These monomers are incorporated into a copolymer to provide cure sites. The relative amounts of these functional monomers present in the polymer chain determines the cross link density which can be achieved on final cure. In this way they have a definite bearing on the film performance properties such as chemical resistance, scratch resistance, flexibility, etc. However, they are usually only minor components (typically acrylamide is used up to levels of $8-10\%$ of the total monomer composition) and the remaining monomer types forming the backbone have a very great influence on the film properties. It should be noted that when acrylamide is used in a polymer it is necessary to methylolate the amide group before a useful cross link site can be achieved. This is usually performed in-situ after the solution polymer has been produced (see chapter on Phenolic and Amino Resins). However, pre-methylolated acrylamides are commercially available and are sometimes used as monomers to avoid this second reaction stage of processing.

Thermoplastic polymer types do not contain functional groups to aid film formation and hence the performance characteristics are solely determined by the monomer types employed in the 'backbone' chain, and by the molecular weight and molecular weight distribution of the polymer.

INITIATORS

The initiators employed are exclusively of the free radical type and are usually organic peroxides or azo compounds. Free radical generation is by thermal decomposition of the initiator (Redox systems are rarely used although the presence of chain transfer agents such as mercaptans will accelerate decomposition) and therefore the half life stability of the initiator at various temperatures is an important choice in deciding exactly which initiator to use at a given process temperature. Initiators and half lives were considered in detail at the beginning of this chapter.

The choice of solvent in a reflux reaction determines the temperature and this in turn influences the choice of initiator. Initiator activity is often expressed as its 'half life' at a specified temperature. A half life of $1-4$ hours is preferred for both commercial and academic work. Azoisobutyronitrile, AIBN or AZDN, is often used in preference to peroxy initiators which can cause oxidation and greater degrees of graft copolymerisation, particularly with acrylates.

CHAIN TRANSFER AGENTS (Modifiers)

These are frequently used, particularly in the polymerisation of thermosetting polymers, to control the molecular weight and hence the solution viscosity of the product. By far the most effective readily available transfer agents for solution polymerisation are mercaptans and these are extensively used commercially. In particular, t-dodecyl mercaptan is often used. Typical levels of mercaptan would be 0.1 to 0.5% of monomer content. One of the disadvantages of the use of mercaptans is the residual odour they impart. This is a particular disadvantage where the polymer is intended for use in connection with food packaging. In these cases alternative methods of reducing the molecular weight must be used (i.e. increase in reaction temperature, use of a shorter half life peroxide, etc.).

SOLVENT

If the polymer is sold in solution form then the solvent(s) used must be acceptable for the end use. This consideration limits both the suppliers selection of solvents and the customer's flexibility in formulation. Solvent selection is considered in more detail later.

The solvent performs a dual role. It acts as a heat sink for the reaction and also acts to dissolve the polymer produced, so maintaining the viscosity of the reaction mixture low enough to allow effective agitation by the reactor stirrer.

The quantity of solvent present varies with the monomer type and intended use of the polymer, but is normally about 40% of the total reactor charge and is rarely less than 30%. At this latter level the viscosity of the reactants is approaching the maximum allowable for efficient reaction agitation. The solvent type is dictated by the polymer end use and solubility factors. Frequently the solvents used are aromatic hydrocarbons (e.g. xylene) or esters (e.g. butyl acetate) or even higher boiling alcohols and ketones (e.g. cellosolves, methyl ethyl ketone).

When selecting solvents the transfer potential of the solvent must also be taken into account. Although the transfer constants of even the most polar solvents are relatively low compared to monomers, polymer and the more conventional chain transfer agents, the high concentration of the solvent present in solution polymerisation increases the chance of termination by transfer to solvent.

THE TROMSDORF EFFECT

This description is valid for systems in which there is free movement of all species in solution. As polymerisation proceeds the solution becomes progressively more viscous and the movement of all molecules is reduced, the larger molecules being those most affected. In these circumstances polymer radicals can no longer freely contact each other and the rate of termination is greatly reduced. The number of radicals increases and the rate of reaction increases correspondingly. At the same time the life of a polymer radical is increased so its molecular weight also increases. This is termed the Tromsdorf effect and is important in all commercial polymerisation where high conversion of monomer into polymer is required. It has been referred to as the gel effect and in recent years it has also been called the Tromsdorf-Norrish effect.

ADVANTAGES OF THE SOLUTION POLYMERISATION PROCESS
  i) Controlled reaction temperatures leading to a more even molecular weight distribution.
  ii) Reproducible products.

DISADVANTAGES
  i) Low molecular weight polymers that do not form highly resistant films unless cross linking (i.e. 'curing') is used.
  ii) Polymer is in solution in an organic solvent which is usually aromatic or highly polar in nature.

The factors affecting molecular weight and solution viscosity are summarised as follows.

FACTORS AFFECTING MOLECULAR
WEIGHT AND SOLUTION VISCOSITY
  i) **Reaction Temperature**
  Generally the molecular weight of the polymer decreases with increasing temperature of reaction. More chains are initiated, therefore the chains are shorter and have lower molecular weight and viscosity. A close control of temperature, which is easier to achieve with refluxing solvent, ensures more consistent molecular weights.

  ii) **Monomer Concentration**
  The molecular weight is influenced by the monomer concentration, usually the lower the concentration of monomers (lower final non-volatile content) the lower the molecular weight of the polymer formed. High concentrations of monomer generate a greater chance that the monomers will collide and, therefore, a reduced possibility of chain termination by solvent contact. Hence the resulting polymer will have a higher molecular weight and higher viscosity.

iii) **Solvent**

The monomers should preferably be soluble in the solvent being used, to obtain a consistent solution acrylic. Variation in solution viscosity for a given monomer mixture depends upon the solvents selected. Different degrees of chain transfer reactivity associated with different solvents give different molecular weights.

The solvents with the highest chain transfer constant give polymers with the lowest molecular weight. The peroxide reactivity (half life) is also influenced by the solvents.

iv) **Concentration of Initiator**

The weight of initiator used can vary between about 0.2−4%. When the concentration of initiator is reduced the molecular weight of the polymer is increased but the conversion of monomer to polymer is reduced. Higher concentrations of initiator form more polymer chains. These chains are, therefore, shorter and hence molecular weight and viscosity are lower.

v) **Polymerisation Regulators (Chain Transfer Agents)**

The chain transfer agent terminates a growing chain whilst creating a new radical capable of initiating a further chain. Chain transfer agents are often used when high temperatures, high concentrations of initiator, and high monomer concentrations are impracticable methods of producing lower molecular weights or viscosities. The higher the concentration of the chain transfer agent the lower the molecular weight and hence the lower the solution viscosity.

SOLUTION POLYMERISATION
— PLANT AND PROCESS

Solution polymerisation, is the most common method for producing acrylic resins for the coatings industry. The main advantage of the diluent or solvent is to take up the heat generated by the polymerisation reaction. It is usual to carry out the reaction at the reflux temperature of the solvents as this allows better control of the reaction temperature.

Plant

Typical plant for the manufacture of acrylic resins consist of a large stainless steel reactor of capacity 5000−15,000+ litres. The reactor is usually heated by steam by means of an external jacket. Temperatures up to 130°C are sometimes required. Cooling is achieved by passing cold water through the same jacket. With reactors of larger capacities it may be necessary to have internal cooling coils to ensure safe control of the large quantities of heat generated by the exothermic reactions, and heating is via zoned, externally wound steam coils.

The reactor's contents are stirred by means of a multi-bladed agitator or turbine, preferably driven by a variable speed motor. The reactor is equipped with a reflux condenser, water separator (for reactions involving water removal) and a receiver. The reactor lid should have light and sight glasses, inlets for monomer and solvent addition, as well as a bursting disc, suitably routed to a dump tank.

It is preferable to discharge the resin through a bottom outlet valve from the reactor into a blender, so that adjustments and filtration may be carried out without causing delays in the manufacturing programme.

Process

The manufacturing process usually consists of a controlled addition of monomers and initiator to refluxing solvent in the reactor. The monomers are fed into the reactor from a weigh (measuring) tank either by a gravity feed via a flow meter, or else pumped using a metering pump. The peroxide may be either premixed with the monomers or added at a controlled rate from a smaller separate tank. The additions of monomers and initiators are carried out at a controlled rate over periods varying from one to six hours. Further additions of peroxide (booster shots or spikes) may be necessary to fully convert the monomers to polymer and ensure a low level of free monomer.

Usually solvents such as aromatic hydrocarbons, esters, or ketones are used for polyacrylates and polymethacrylates. Higher alkyl polymers are soluble in aliphatic hydrocarbons, alcohols or ethers. Normally the reaction is carried out in the quantity of solvent required to ensure that the final non-volatile content is 50−70%. Sometimes the reaction is carried out at higher monomer concentrations and then diluted after polymerisation is complete in order to achieve higher molecular weight polymers.

The amounts of initiators used vary between 0.2% and 4% and may be any of the types described earlier, depending upon the monomers and solvents.

The total processing time varies between 8−30 hours, the reaction end point being the conversion of greater than 99% of monomers to polymer, determined by non-volatile content. Also, the viscosity is measured to ensure the required molecular weight has been obtained.

The majority of monomer vapours will form explosive mixtures with air. Thus, if a reactor is charged and heated to reaction temperature, then in all probability the contents will pass through an explosive mixture. The use of nitrogen prior to charging and a gentle flow of nitrogen during processing can eliminate the ingress of air. The exclusion of oxygen is desirable to avoid potential inhibition of polymerisation.

TYPES OF SOLUTION VINYL RESINS

As already stated vinyl (and acrylic) resins can be divided into thermoplastic and thermoset types and they are both considered in the following sections. For completeness, dispersion acrylics are briefly mentioned.

Thermoplastic Solution Resins

Thermoplastic coatings or inks contain resin, solvent and probably other ingredients (pigments, etc.). A tack free film must be formed upon evaporation of the solvent. Thus, the resins used for these coatings must be 'film formers' in their own right. Their Tg's must be sufficiently low to film form but not so low that they block or cold flow. Thus, the commonly used thermoplastic vinyl and acrylic resins are restricted to those formed from monomers or co-monomers in ratios to give the desired film properties. The monomers must also be selected to give performance, protection and resistance to the environment of the coating. Because these resins do not cross-link, all of the desired properties must be incorporated by careful monomer selection in combination with economic considerations.

This should explain to an outsider to the resin industry, why so many thermoplastic resins are derived from a relatively few monomers. Unfortunately, a realistic list of usable thermoplastic monomers is severely limited.

Large quantities of acrylic lacquers are used in finishes and refinishes for automobiles. They give good reflow properties which are required for the bake-sand-bake techniques of application. This allows small imperfections in the coating to be sanded out and then the coating to be rebaked at 140°C so that reflow occurs, resulting in a high gloss finish. These types of coating are based on methyl methacrylate, externally plasticated with phthalates or internally plasticised by co-polymerisation with a lower Tg acrylate monomer. The molecular weight of this type of polymer has to be high to obtain good gloss retention during exterior exposure. However, there is an upper limit for the molecular weight of 100,000 to 110,000, above which cobwebbing is likely to occur during the spraying of the coating. Therefore, the molecular weight distribution has to be carefully controlled.

Methyl methacrylate can be polymerised in a solvent mixture of toluene and acetone using benzoyl peroxide or azodisobutyronitrile as free radical initiators. The monomer containing $\frac{1}{2}\%$ initiator can be gradually added to the solvent at reflux over a period of several hours, but a narrower distribution of molecular weight can be obtained by adding all the reactants at the same time and heating to 110°C under pressure.

The principal advantages of thermoplastic acrylic coatings are their clarity and low colour coupled with their outstanding exterior durability arising from their resistance to hydrolysis and ultra-violet degradation. In addition to their use in original finishes for cars, they are used in car repair finishes. Other major uses are as clear sealers for concrete roof tiles and floors, and as coatings and inks for plastics, films and foils.

Polystyrene, unlike acrylic resins, yellows badly during prolonged exterior exposure, but some styrene monomer is often mixed with the acrylic monomer before polymerisation to reduce raw material costs. Odour of acrylic resins can be a major disadvantage in ink applications.

Vinyl chloride is a relatively inexpensive and toxic monomer which gives polymers of good colour and chemical resistance. Poly (vinyl chloride) copolymer resins for coatings are often referred to simply as 'vinyl resins' and are frequently used in coating lacquers for food and beverage cans, and for foil and wire coatings.

Vinyl chloride is usually polymerised by the suspension method (see later section). A co-monomer is often included to improve the solubility and adhesive properties and to act as an internal plasticiser. However, external plasticisers of the tricresyl phosphate or the phthalate type are normally included in the coating formulation to help reduce solvent retention. A typical monomer mix would be vinyl chloride, vinyl acetate and methacrylic acid in the ratio 80:19:1.

Vinyl chloride polymers are soluble in ketones but co-polymers are often soluble in ketone-hydrocarbon mixtures.

Vinyl chloride polymers can undergo degradation by the elimination of hydrogen chloride if exposed to temperature above about 180°C or to prolonged u.v. light or an oxidising environment:

$$- CH_2 - CH - CH_2 - CH \qquad \xrightarrow{(-HCl)} \qquad - CH_2 - CH - CH = CH -$$
$$\qquad\qquad |\qquad\quad\ |\qquad\qquad\qquad\qquad\qquad\qquad\qquad\qquad |$$
$$\qquad\qquad Cl\qquad\quad Cl\qquad\qquad\qquad\qquad\qquad\qquad\qquad\qquad Cl$$

The eliminated hydrogen chloride then catalyses further dehydrochlorination and the polymer degrades and discolours. It is therefore advisable to include a HCl acceptor to act as a stabiliser.

The actual mechanism of dehydrochlorination is subject to much debate and dis-agreement and there are many discrepancies between theory and practice. Stabilisers for p.v.c. include organo tin complexes and soaps of heavy metals like lead or cadmium. Soaps of calcium and barium are also becoming more popular to replace the more toxic metals.

Thermosetting Solution Coatings

Many high performance coatings are formulated from solution vinyl or acrylic resins which react after being applied to the substrate and after the solvent has evaporated. The reaction is often brought about by heat, i.e. by stoving, but other curing processes are also common, e.g. isocyanate reaction.

The resins are made by similar processes to the thermoplastic acrylic resins (lacquer-drying acrylics) discussed in the previous section. However, so-called functional monomers are included in the monomer mix so that pendent functional groups are present along the polymer chain after polymerisation.

The chain length of thermosetting resins is extended during the cross-linking reaction and therefore the molecular weight of the uncured resin can be much lower than that of a thermoplastic resin. Therefore, the lower molecular weight thermosetting types are easier to dissolve and can be used to formulate high solid coatings. The resin content of thermoplastic coatings based on poly (methyl methacrylate) has to be as low as 25% when the molecular weight is as high as 100,000. Also, the better solubility properties of the lower molecular weight thermosetting polymers means that they can be dissolved in a wider range of solvents, and so there is more scope for cost reductions, etc.

Another major advantage of thermosetting resins is that cross-linked films can give better gloss, chemical and solvent resistance, and blocking resistance when their films are at ambient or higher temperatures.

Thermosetting acrylics can be classified according to their reactive groups, i.e. carboxyl, hydroxyl, anhydride, epoxide, amine, isocyanate, acrylamide or N-methylol ether. These polymers can be made in one stage by including in the monomer mixture a monomer with the required functionality, e.g.

$$CH_2 = CH - CO_2 - CH_3 \quad + \quad CH_2 = CH - CO_2 - H \quad \longrightarrow$$

Methyl acrylate $\qquad\qquad\qquad\qquad$ Acrylic acid $\qquad\qquad\qquad\qquad$ $\underset{|}{\overset{}{}}$ CO_2H

Alternatively, the required functional group can be formed by modifying the polymer after polymerisation, e.g.

$$\text{~~~} \underset{CO_2H}{|} \quad + \quad \overset{O}{\overset{/\backslash}{CH_2 - CH - CH_3}} \quad \longrightarrow \quad \text{~~~} \underset{\underset{OH}{\overset{|}{OCH - CH - CH_3}}}{\overset{|}{\underset{|}{CO}}}$$

Propylene oxide

One of the most important examples of this second type are the acrylamide copolymers which are post-reacted with formaldehyde and an alcohol to form the N-alkoxy methyl ether of the amide, and are considered in detail later.

Some examples of the range of available functional monomers is given below.

RANGE OF FUNCTIONAL MONOMERS

Functional Group	Monomer
Carboxyl	Acrylic acid Methacrylic acid
Hydroxyl	Hydroxyethyl acrylate and methacrylate Hydroxypropyl acrylate and methacrylate
Anhydride	Maleic anhydride Itaconic anhydride
Epoxide	Glycidyl acrylate and methacrylate Alkyl glycidyl ether
Amine	Dimethylaminoethyl methacrylate
Isocyanate	Vinyl and allyl isocyanate
Amide	Acrylamide Methacrylamide Maleimide

The inclusion of the functionality can be considered to be incorporated by either internally reactive groups or potentially reactive ones, and it is the latter which play a major role in surface coating chemistry.

INTERNALLY REACTIVE GROUPS
The following groups may be incorporated onto the acrylic backbone to provide a self-reactive system.

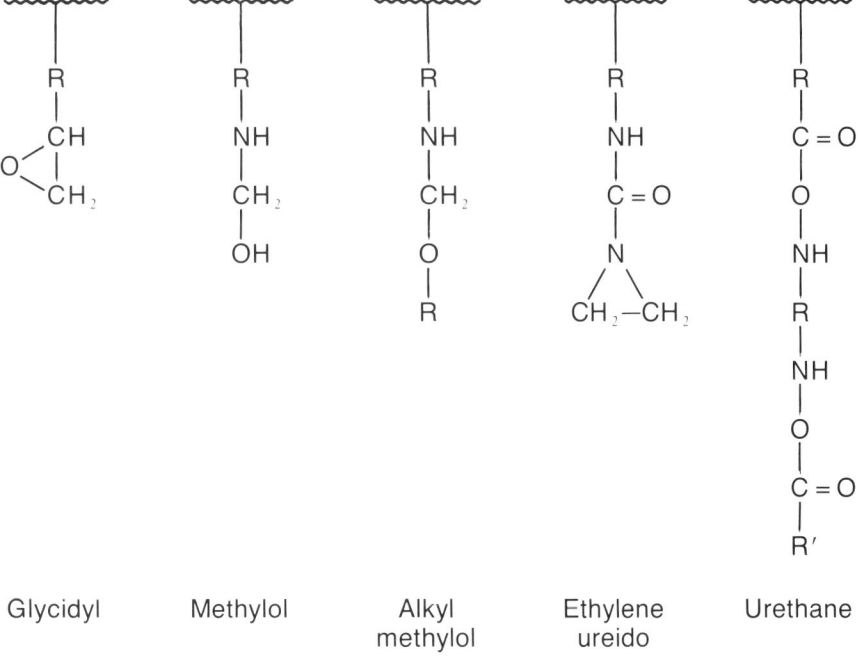

| Glycidyl | Methylol | Alkyl methylol | Ethylene ureido | Urethane |

Other co-reactive groups such as carboxyl, hydroxyl, epoxy, isocyanate, methylol or amine may be used with systems of this type. Catalysts, usually acids or amines are necessary to produce the required degree of cure, this unfortunately tends to result in systems with poor shelf life.

POTENTIALLY REACTIVE GROUPS

The following functional groups may be built into the acrylic polymer to provide potential sites for cross-linking.

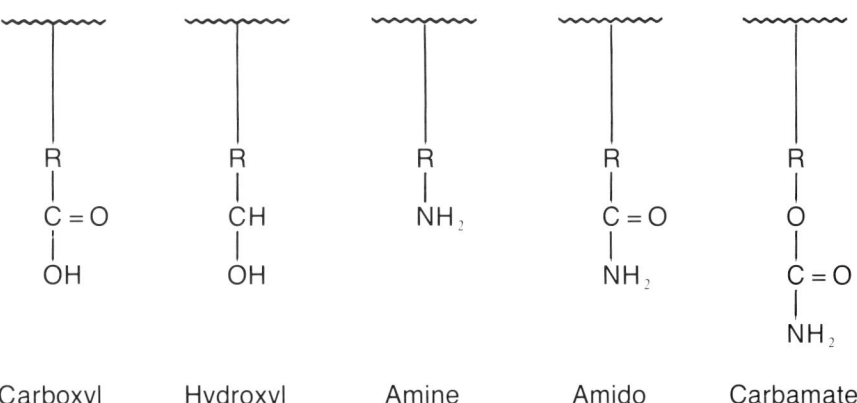

| Carboxyl | Hydroxyl | Amine | Amido | Carbamate |

These may be cross-linked with:

Carboxyl Group –
Epoxy resins, oxirane triazines, polyvalent metal salts.

Hydroxyl Groups –
Amino formaldehyde resins, isocyanates.

Amine Group –
Epoxy resins, isocyanates, aldehydes, amino formaldehyde resins.

Amide/Carbamate Group –
Aldehydes.

Only four of these groups have been introduced into acrylic resins, commercially. These are the hydroxyl, carboxyl, epoxy and alkyl or alkyl methylol groups, although some commercially available resins contain amine comonomers at relatively low levels (typical 2%, normally less than 5%), to give specialist properties.

Some of the cross-linking reactions can be summarised as follows in generalised and simplified representations.

SIMPLIFIED REACTION STRUCTURES

Hydroxyl containing Acrylic Resins

WITH AMINO FORMALDEHYDE RESINS

ii)

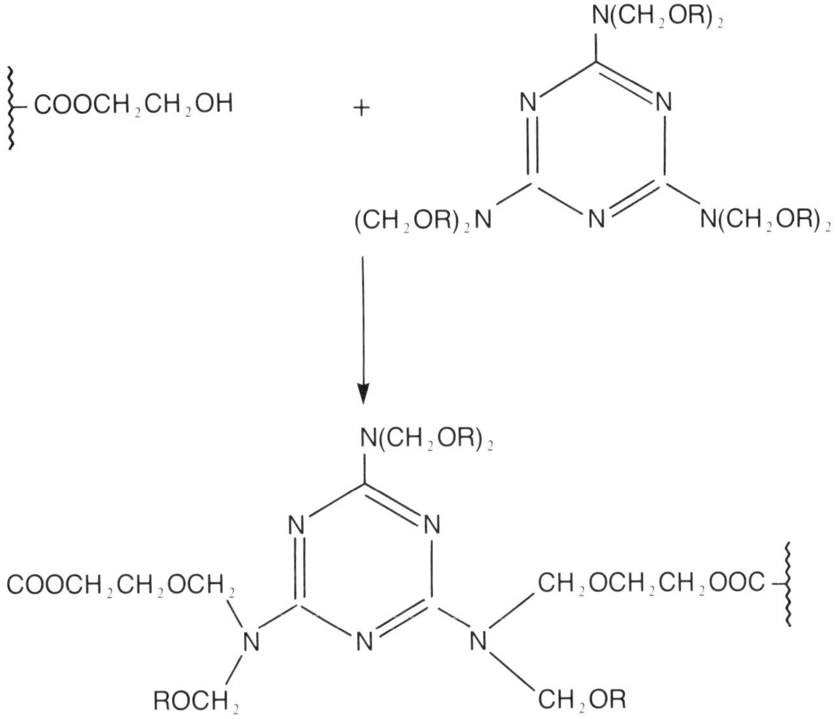

$$\{-COOCH_2CH_2OH \quad + \quad \text{[triazine: } N(CH_2OR)_2 \text{ substituted s-triazine ring with } (CH_2OR)_2N \text{ and } N(CH_2OR)_2 \text{]}$$

↓

$$\{-COOCH_2CH_2OCH_2 \text{—[triazine ring with } N(CH_2OR)_2 \text{ top, } N\text{-}ROCH_2 \text{ and } N\text{-}CH_2OR \text{]—} CH_2OCH_2CH_2OOC-\}$$

iii)

$$\{-NH-CH_2-OR \;+\; HO-\} \quad \longrightarrow \quad \{-NH-CH_2-O-\} \;+\; ROH$$

WITH ISOCYANATES (usually two pack systems)

$$\{-COOCH_2CH_2OH \;+\; OCN-R-NCO \;+\; OHCH_2CH_2OOC-\}$$

↓

$$\{-COOCH_2CH_2OOCHN-R-NHCOOCH_2CH_2OOC-\}$$

For one pack systems blocked isocyanate can be used (see Polyurethane chapter in Volume III).

182

WITH POLYANHYDRIDES

$$\{-COOCH_2CH_2OH \quad + \quad C \quad C \quad C \quad C \quad + \quad OHCH_2CH_2OOC-\}$$

$$\{-COOCH_2CH_2OOC \quad C \quad C \quad COOCH_2CH_2OOC-\}$$

WITH EPOXY RESINS (with acid catalysts and high stoving temps)

$$\{-COOCH_2CH_2OH \quad + \quad CH_2 - -CH - CH_2 - O - \langle \rangle - \underset{CH_3}{\overset{CH_3}{C}} - \langle \rangle -$$

$$\{-COOCH_2CH_2OCH_2CH - CH_2O - \langle \rangle - \underset{CH_3}{\overset{CH_3}{C}} - \langle \rangle -$$
$$\qquad\qquad\qquad\quad OH$$

WITH CARBOXYL GROUPS

$$\{-CO_2H \quad + \quad HO-R-\} \quad \longrightarrow \quad \{-\overset{O}{\overset{\|}{C}}-O-R-\} \quad + \quad H_2O$$

High temperatures and catalysts are required.

Carboxyl containing Acrylic Resins

WITH EPOXY GROUPS

Under certain conditions further reaction through the hydroxyl group is possible.

WITH ISOCYANATE GROUPS

$$2 \quad \{-CO_2H \ + \ OCN-R-NCO \ \longrightarrow \ \{-\overset{O}{\overset{\|}{C}}-\overset{H}{\overset{|}{N}}-R-\overset{H}{\overset{|}{N}}-\overset{O}{\overset{\|}{C}}-\} \ + \ 2CO_2$$

Care must be exercised to minimise CO_2 evolution.

WITH ALKOXYL METHYL ETHER GROUPS

$$\xi\text{—}CO_2H \ + \ RO\text{—}CH_2\text{—}NH\text{—}\xi \ \longrightarrow \ \xi\text{—}\overset{\displaystyle O}{\overset{\displaystyle \|}{C}}\text{—}O\text{—}CH_2\text{—}NH\text{—}\xi \ + \ ROH$$

The reactant noted as polymer$-$NH$-$CH$_2$OR can be the N-alkoxy methyl ether of an acrylamide polymer or the alkoxy methyl ether of a melamine or urea formaldehyde resin.

Glycidyl containing Acrylic Resins

WITH EPOXY GROUPS

Self reaction can occur.

WITH AMINE GROUPS

Note: R can be H.

WITH AMINE FORMALDEHYDE RESINS

WITH PHENOLIC RESINS

Methylol containing Acrylic Resins

These acrylic resins contain pendant amide groups introduced into the polymer by the use of acrylamide or methacrylamide. One method of producing methylol groups in the acrylic chain is to react the amide containing polymer with formaldehyde to produce methylol groups and then further etherify these groups with butyl alcohol under acid conditions.

$$\underset{\substack{\displaystyle | \\ C=O \\ | \\ NH_2}}{\sim\!\!\sim\!\!\sim C\sim\!\!\sim\!\!\sim\!\!\sim\!\!\sim\!\!\sim\!\!\sim\!\!\sim\!\!\sim\!\!\sim} \quad \underset{\substack{\displaystyle | \\ C=O \\ | \\ NH_2}}{C\sim\!\!\sim\!\!\sim} \qquad + \ 2 \ HCHO$$

$$\downarrow$$

$$\underset{\substack{\displaystyle | \\ C=O \\ | \\ NH \\ | \\ CH_2OH}}{\sim\!\!\sim\!\!\sim C\sim\!\!\sim\!\!\sim\!\!\sim\!\!\sim\!\!\sim\!\!\sim\!\!\sim\!\!\sim\!\!\sim} \quad \underset{\substack{\displaystyle | \\ C=O \\ | \\ NH \\ | \\ CH_2OH}}{C\sim\!\!\sim\!\!\sim} \qquad + \ 2 \ BuOH$$

$$\downarrow$$

$$\underset{\substack{\displaystyle | \\ C=O \\ | \\ NH \\ | \\ CH_2OBu}}{\sim\!\!\sim\!\!\sim C\sim\!\!\sim\!\!\sim\!\!\sim\!\!\sim\!\!\sim\!\!\sim\!\!\sim\!\!\sim} \quad \underset{\substack{\displaystyle | \\ C=O \\ | \\ NH \\ | \\ CH_2OBu}}{C\sim\!\!\sim\!\!\sim}$$

These methylol or butoxy methyl groups can react with themselves, epoxy, carboxyl, hydroxy or amino formaldehyde resins.

SELF CONDENSATION

i) CONDENSATION

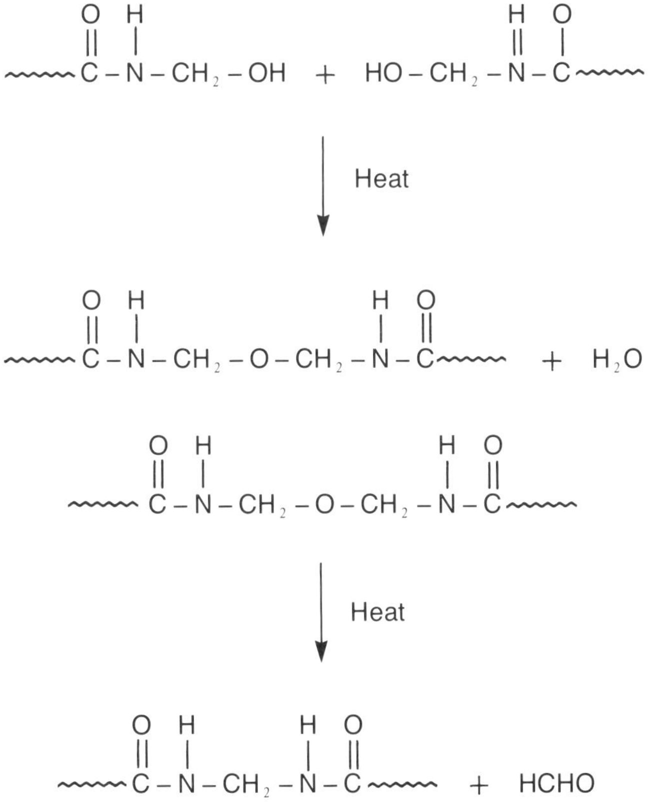

$$\text{\sim\sim}\overset{\overset{\displaystyle O}{\|}}{C}-\overset{\overset{\displaystyle H}{|}}{N}-CH_2-OH \quad + \quad HO-CH_2-\overset{\overset{\displaystyle H}{|}}{N}-\overset{\overset{\displaystyle O}{\|}}{C}\text{\sim\sim}$$

Heat

$$\text{\sim\sim}\overset{\overset{\displaystyle O}{\|}}{C}-\overset{\overset{\displaystyle H}{|}}{N}-CH_2-O-CH_2-\overset{\overset{\displaystyle H}{|}}{N}-\overset{\overset{\displaystyle O}{\|}}{C}\text{\sim\sim} \quad + \quad H_2O$$

$$\text{\sim\sim}\overset{\overset{\displaystyle O}{\|}}{C}-\overset{\overset{\displaystyle H}{|}}{N}-CH_2-O-CH_2-\overset{\overset{\displaystyle H}{|}}{N}-\overset{\overset{\displaystyle O}{\|}}{C}\text{\sim\sim}$$

Heat

$$\text{\sim\sim}\overset{\overset{\displaystyle O}{\|}}{C}-\overset{\overset{\displaystyle H}{|}}{N}-CH_2-\overset{\overset{\displaystyle H}{|}}{N}-\overset{\overset{\displaystyle O}{\|}}{C}\text{\sim\sim} \quad + \quad HCHO$$

ii) CONDENSATION

$$\text{\sim\sim}\overset{\overset{\displaystyle O}{\|}}{C}-\overset{\overset{\displaystyle H}{|}}{N}-CH_2-OH \quad + \quad BuO-CH_2-\overset{\overset{\displaystyle H}{|}}{N}-\overset{\overset{\displaystyle O}{\|}}{C}\text{\sim\sim}$$

Heat

$$\text{\sim\sim}\overset{\overset{\displaystyle O}{\|}}{C}-\overset{\overset{\displaystyle H}{|}}{N}-CH_2-O-CH_2-\overset{\overset{\displaystyle H}{|}}{N}-\overset{\overset{\displaystyle O}{\|}}{C}\text{\sim\sim} \quad + \quad BuOH$$

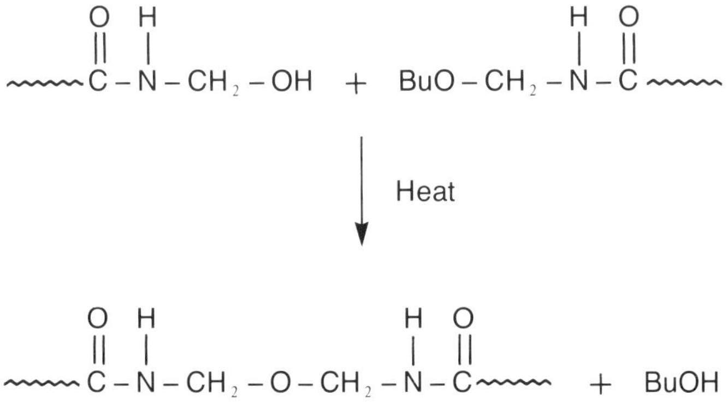

iii) CONDENSATION

$$\text{\textasciitilde\textasciitilde\textasciitilde} \underset{\overset{||}{O}}{C} - \underset{\overset{|}{H}}{N} - CH_2 - OBu \quad + \quad BuO - CH_2 - \underset{\overset{|}{H}}{N} - \underset{\overset{||}{O}}{C} \text{\textasciitilde\textasciitilde\textasciitilde}$$

↓ Heat

$$\text{\textasciitilde\textasciitilde\textasciitilde} \underset{\overset{||}{O}}{C} - \underset{\overset{|}{H}}{N} - CH_2 - \underset{\overset{|}{H}}{N} - \underset{\overset{||}{O}}{C} \text{\textasciitilde\textasciitilde\textasciitilde} \quad + \quad BuO - CH_2 - OBu$$

Considering the N-methylol polymer only for simplicity (the N-methylol ether will follow similar reaction schemes). Epoxy, carboxyl, hydroxyl schemes have already been covered but are repeated here for completeness and because the reactions of the N-methylol group (or ether) are a basis for a major portion of the thermosetting vinyl resins which play a prominent industrial and commercial role.

SELF CONDENSATION

$$2 \;\{NH - CH_2 - OR \quad \longrightarrow \quad \{NH - CH_2 - NH\} \quad + \quad 2HCHO$$

EPOXY GROUP

$$\{NHCH_2 - OR \; + \; CH_2 - CH \text{\textasciitilde\textasciitilde} \quad \longrightarrow \quad \{NHCH_2 - O - CH_2 - CH \text{\textasciitilde\textasciitilde}$$
$$\overset{| \qquad \qquad}{OR}$$

CARBOXYL GROUP

$$\{NHCH_2 - OR \; + \; HO_2C\} \quad \longrightarrow \quad \{NHCH_2 - O - \underset{}{\overset{\overset{O}{||}}{C}} \text{\textasciitilde\textasciitilde}\} \quad + \quad ROH$$

HYDROXYL GROUPS

$$\text{\}-\text{NHCH}_2-\text{OR} \quad + \quad \text{HO}-\text{\{} \quad \longrightarrow \quad \text{\}-\text{NHCH}_2-\text{O}-\text{\{} \quad + \quad \text{ROH}$$

AMINE GROUPS

$$\text{\}-\text{NHCH}_2\text{OR} \quad + \quad \text{H}_2\text{N}-\text{\{} \quad \longrightarrow \quad \text{\}-\text{NH}-\text{CH}_2-\text{NH}-\text{\{} \quad + \quad \text{ROH}$$

Uses of Thermosetting Vinyl Resins

The acrylamide modified polymers give tough, resistant films which are used for washing machines, refrigerators and white goods in general.

Carboxyl function polymers, usually with epoxy resins tend to be used for domestic appliances and metal decorating applications.

The hydroxyl modified thermosetting acrylic cured with amino resins find wide application in automotive finishing and these coatings generally have the best exterior durability. The majority of these coatings generally have good exterior durability, and are one pack systems requiring stoving at 120−160°C for 30 minutes to produce an adequately cured system. Two pack hydroxyl cured systems mainly use isocyanates as the cross-linking agent.

SOLVENT SELECTION

The principal factors affecting selection of solvents for a specific polymer finish are cost, solvent power for the polymer, evaporation rate, flammability, toxicity and customer requirements. Costs vary with time and quantity used, but hydrocarbons, alcohols and ketones produced by continuous petrochemical processes are relatively cheap; these include ethanol, isopropanol, isobutanol, acetone, methyl ethyl ketone and methyl isobutyl ketone. Acetate esters of these alcohols and chlorinated solvents are rather more expensive while other solvents may be regarded as speciality chemicals and can be very costly.

The viscosity of polymers is lowest in good solvents and highest in poor solvents; there is a synergistic effect with aromatic or chloro solvents with esters which makes both groups of liquids better solvents than their solubility parameter would suggest. For the same reason, a mixture of two solvents of different groups is often better than a single solvent. Despite a lot of work, solvent selection is still more of an art than a science and experience is the most significant contributor.

The rate of evaporation depends mainly on the vapour pressure of the solvent which decreases as the boiling point rises.

Solvents generally evaporate more slowly than their vapour pressure would suggest, but this effect is not of major significance for assessing possible solvency.

The flammability of a solvent is normally measured by its flash point which is defined as the temperature at which the mixture of air and water above the solvent (or solution) will just burn. It is normally measured in a closed cup apparatus; some values are shown below. The values shown are in degrees Celsius but it should be noted that many reference books quote flash points in degrees Fahrenheit.

**FLASH POINTS AND BOILING POINTS
OF SOME COMMON SOLVENTS**

Solvent	Boiling Point °C	Flash Point °C
Di-ethyl ether	34	− 29
Hexane	69	− 22
Acetone	56	− 18
Cyclohexane	81	− 17
Benzene	80	− 11
Heptane	98	− 4
Ethyl acetate	77	− 4
Methyl ethyl ketone	80	− 1
Toluene	111	+ 4
Di-chloroethane	84	6
Methanol	65	12
Ethanol	78	13
Octane	126	13
o-Xylene	144	18
Butanol	118	22
Butyl acetate	126	22
Amyl acetate	141	24
Chlorobenzene	132	32
Cellosolve	156	40

The flash point indicates how readily a solvent will ignite. Paintmakers recommendations are: Use CO_2, dry chemical, sand or earth for small fires; for larger fires use alcohol type foam or water spray. Acetone and the lower alcohols are miscible with water which will extinguish their fires. Hydrocarbons and esters are both immiscible with water and lighter than it. Consequently they will float on any water pumped onto the fire. For large fires this will result in spreading the blaze. A few solvents such as di-chloroethane are heavier than water and may be extinguished by excluding air.

Although it is an excellent solvent for many polymers, benzene is not used industrially due to its high toxicity; among other effects it damages red blood cells and causes septic pneumonia. Many chloro solvents are toxic and only 1,1,1 tri-chloro ethane (Methyl chloroform or Genklene) is used to any appreciable extent. Di-oxan and nitrobenzene are among the speciality solvents whose use has been discontinued on grounds of toxicity.

Water is the ideal solvent from the cost and pollution viewpoints but it is a non-solvent for many surface coating polymers. It will dissolve a small number of homopolymers notably those derived from acrylamide, acrylic acid, itaconic acid, vinyl methyl ether, vinyl pyrrolidone and vinyl sulphonic acid, but none of these homopolymers forms flexible films of use in the coatings industry. While copolymers of acrylic or methacrylic acids with acrylate esters are generally insoluble in water, their salts are soluble when the acid content is over 5% (for hydrophilic monomers) and 12% (for hydrophobic monomers). Such polymers can be prepared in solution, or in emulsion, but not in aqueous solution. This is because the acrylate esters are insoluble in water. The acid is copolymerised in the un-ionised form because the ion is unreactive to free radicals. In emulsion polymerisation care has to be taken to avoid homopolymerisation of the acrylic or methacrylic acid in the water phase. Suppression of homopolymerisation requires a low concentration of acid throughout the polymerisation process. This can be achieved by using a long reaction period and slow addition of monomer mixture, or by careful pH buffer selection.

Sodium, potassium and ammonium salts will all solubilise acidic copolymers but, in practice, ammonium salts are used because the ammonia evaporates slowly from the polymer film. After ageing, the polymer film reverts to the un-ionised acid form and is insoluble in water with a lower water sensitivity than the salt form. Another method of reducing water sensitivity is to add certain heavy metal salts of which zinc and zirconium salts are the most important. This technology will be discussed more fully in the section on floor polishes.

The drying of coatings based on polymer solutions is designed to give a suitable open time to allow good application. When a mixture of solvents is used a change in solvent composition will occur during drying. Should this change result in partial precipitation of polymer film, faults such as blooming and blushing can occur. In formulating solvent blends, care must be taken to ensure that the polymer will remain in solution throughout the drying process. To ensure that this occurs, the 'best' solvent should have the highest boiling point of the mixture. Relative humidity produces problems for both aqueous and non-aqueous solutions. With solvent finishes, solvent evaporation will cause a drop in temperature at the surface, and at high humidities and high rates of evaporation this may result in the surface temperature becoming lower than the dew point of the air, where-upon water will condense on the surface causing local precipitation of polymer. In the case of aqueous solutions, the rate of evaporation is proportional to the dryness of the air or 100 minus relative humidity. At very high humidities water based coatings will not dry at all.

Solvent power is a very complex subject and many resins and coatings formulators use the solubility parameter approach. The majority of the polymer solution theories are beyond the scope of this book and only relate to academic systems which utilise low concentrations of polymer.

Solubility and Solubility Parameters

In classical polymer lattice theories, Huggins, *et al*, showed that the free energy of mixing a solvent with a linear homogenous polymer was related to:

i) The molar volume of the solvent.

ii) The molar volume of a segment (monomer unit) of the polymer.

iii) The volume fractions (which is a way of expressing concentration) of both polymer and solvent.

iv) RT (gas constant and temperature in degrees absolute).

v) An empirical constant dependent upon the co-ordination number of the lattice.

vi) A heat of mixing constant which is specific for that particular mixture.

$$\Delta H = K\phi_s \phi_p$$

Where, ΔH is the enthalpy (heat) of mixing, k is a constant. ϕ_s and ϕ_p are volume fractions of solvent and polymer respectively.

Scatchard showed that

$$\Delta H = \phi_s \phi_p [(E_s/V_s)^{\frac{1}{2}} - (E_p/V_p)^{\frac{1}{2}}]^2$$

Where E_s and E_p are molar cohesive energies of solvent and polymer, and V_s and V_p are solvent and polymer molar volumes.

$$K = \Delta H/(\phi_s \phi_p) \qquad \text{from earlier}$$

thus
$$K = [(E_s/V_s)^{\frac{1}{2}} - (E_p/V_p)^{\frac{1}{2}}]^2 = (\delta_s - \delta_p)^2$$

where
$$\delta = \sqrt{E/V}$$

E/V is called the Cohesive Energy Density (C.E.D.) and δ the solubility parameter (Hildebrand).

The free energy of mixing ΔG_m is related to ΔH_m by the equation:

$$\Delta G_m = \Delta H_m - T\Delta S_m$$

For mixing to occur ΔG_m must be negative. As a general rule ΔS_m is positive and much smaller than ΔH_m. Thus, the magnitude of ΔH_m dictates whether solution occurs or not. Thus, as a general rule if ΔH_m approaches zero then solution will occur because $T\Delta S_m$ is the dominant factor. Thus $(\delta_p - \delta_p)^2 = 0$ is the desirable situation. This can be achieved by letting $\delta_s \rightarrow \delta_p$. In other words if the solubility parameter of the solvent equals that of the polymer, the polymer will be soluble. In the early 1950's this approach to estimating polymer solubility was popular and offered a straightforward approach to understanding the thermodynamics of polymer solutions.

For simple molecules like solvents δ can be determined from measurements of heats of vaporisation (or calculated from the Clausius Clapeyron equation).

However, most polymers degrade on vaporisation (if that temperature could be achieved). It is difficult and tedious to measure, or estimate experimentally, δ_p for polymeric molecules.

Experimental methods include determining a polymer's solubility or non-solubility in a range of solvents of known δ_s. This will give a range of solvents of known δ_s in which the polymer is soluble. The value of δ_p is often taken as the mid-point of this range. Solvents of known δ_s must be selected with different chemical structures to avoid other chemical bonding effects influencing the results.

Swelling measurements and vapour pressures can also be used to determine δ_p. Small (Journal Applied Chemistry 1953(3), pps. 71–80) showed that it is possible to estimate solubility parameters of polymers from a knowledge of their chemical structure. He allocated values to each group and then assumed that these values were additive. This approach gave fairly good agreement for hydrocarbon and other relatively non-polar polymers. Uncertainty of polymer density gives errors in estimated solubility. In essence, Small allocated a molar attraction constant for each group (F) and summed these to give δ viz;

$$\delta = \sum F/V$$

Examples of values of F for individual groups are:

$- CH_3$	$214 \ cal^{1/2} \ cm^{3/2}$
$>CH_2$	$133 \ cal^{1/2} \ cm^{3/2}$
$- \overset{\mid}{C}H$	$28 \ cal^{1/2} \ cm^{3/2}$
$- \overset{\mid}{C}H =$	$190 \ cal^{1/2} \ cm^{3/2}$
$-$ Phenyl	$735 \ cal^{1/2} \ cm^{3/2}$
$-$ Cl	$270 \ cal^{1/2} \ cm^{3/2}$

Values reproduced from Small's papers.

The concept of solubility parameters is complicated because volume changes occur on mixing polymer and solvents, and the majority of surface coating resins are polar (to some extent) and many of the solvents have some degree of polarity. The effects of hydrogen bonding, even if only weakly bonded, also affects solubility.

Hydrogen bonding and di-polar effects are the greatest shortcomings of this approach. Other terms have been introduced to compensate for polarity and hydrogen bonding.

Burrell (Official Digest 1955, pps. 726–758) gives lists of solubility parameters of solvents and polymers. In addition, the approach of Small is reviewed.

Another failing of the solubility parameter theory is in the area of solvent mixtures. Two solvents, with δ_s values which would not be expected to dissolve a polymer can at certain mixing ratios dissolve the polymer. An example (given by Burrell) is a solid epoxy resin in xylene and methyl isobutyl carbinol. Theory predicts that as these are both non-solvents then mixtures should also be non-solvents. However, at certain temperatures the epikote resin is dissolved. Solubility parameters are: δ_p 10.9, δ_s 8.8 and 10.0. If the value of δ_p lay between δ_s for both solvents then it would be anticipated that solubility would occur at a ratio corresponding to; their relative proportions; the difference in δ_s; the value of δ_p.

Crystallinity of the polymer can also effect solubility and give unexpected results.

Many film forming polymers have δ_p's in the range 8.5−10. Thus, solvents with δ_s in this range are also preferred. Low molecular weight also tends to increase solubility. The following tables summarise some solubility parameters (taken from the literature) for polymers and solvents.

TYPICAL VALUES OF THE SOLUBILITY PARAMETERS

	Material	Solubility Parameter
Solvents	Acetone	10
	Xylene	8.8
	Toluene	9.0
	Butanol	11.4
	Ethanol	12.8
	Methyl ethyl ketone	9.3
	White spirit	7.0−7.6
	Carbon tetrachloride	8.6
	2-ethoxyethanol	9.9
	Water	23.4
	Propylene glycol	15.0
Polymers	Poly (methyl methacrylate)	9.5
	Poly (ethyl methacrylate)	9.0
	Poly (butyl methacrylate)	8.7
	Poly (styrene)	9.1
	Poly (acrylonitrile)	15.4
	Poly (methyl acrylate)	9.6
	Poly (ethyl acrylate)	9.2
	Poly (butyl acrylate)	8.7
	Poly (vinyl acetate)	9.4
	Poly (vinylidene chloride)	12.2

Material	Solubility Parameter
Polyethylene	7.9
Solid epoxy resin	10.9
Medium oil length alkyd	9.4
Poly (vinyl chloride)	9.7
Ester gum	9.0
Nylon 66	13.6
Phenolic resin	11.5
Rubbers	8.1 – 9.4

It is possible to generally group solubility parameters of solvents for different chemical types.

Solvent Group	Range of Solubility Parameters
Aliphatic hydrocarbons	6.8 – 8.5
Aromatic hydrocarbons	8.5 – 9.5
Chlorinated solvents	8.2 – 10.0
Ethers	7.4 – 9.9
Esters	7.8 – 10.0 (majority)
Ketones	7.8 – 10.4
Alcohols	8.9 – 16.5

As is readily apparent, aromatic hydrocarbons have solubility parameters nearer those of polymers than aliphatic hydrocarbons do. Thus, it is not surprising that aromatic hydrocarbons are generally 'better' solvents for resins than aliphatic ones, even to the extent that some grades of white spirit with a high aromatic content are solvents, whilst other grades with aromatic contents approaching zero are non-solvents, and precipitation or insolubility occurs.

Inspection of solubility parameters gives a very good indication of the more suitable solvent combinations and helps to predict resistance to materials.

Polymers of acrylonitrile are predicted to have good resistance to attack by aliphatic hydrocarbons (e.g. white spirit). These polymers will also be most soluble in solvents with higher solubility parameters (e.g. 2-ethoxyethanol, toluene or butyl alcohol combinations).

Acrylics with larger alkyl chains are more soluble in aliphatic solvents since their solubility parameters are similar. Conversely, they will have good resistance to alcohols, water and glycol because of the large difference in their respective solubility parameters.

An improvement to the original approach to solubility parameters was the separation of the solubility parameter into three parameters each contributing to the overall solubility. These parameters were related to dispersion, polar and hydrogen bonding forces. Some earlier approaches followed these lines but only grouped rather than evaluated the individual contributions.

Three parameters represent the contribution of dispersion, polar and hydrogen bonding forces to the cohesive energy of the molecule. Solvents whose parameters most nearly approach those of the polymer are the best solvent for that polymer. The parameters for some common solvents are shown in the table below (taken from the literature).

SOLUBILITY PARAMETERS FOR SOME COMMON SOLVENTS

Solvent	Dispersion Parameter	Polar Parameter	Hydrogen Bonding Parameter
Hexane	7.3	0	0
Di-ethyl ether	7.3	0.4	1.7
Ethanol	7.7	4.3	9.5
Isopropanol	7.7	3.0	8.0
Isobutanol	7.6	2.8	7.8
Toluene	8.8	0.7	0.2
Cyclohexane	8.2	0.2	0.4
Ethyl acetate	7.6	4.4	1.4
Acetone	7.6	5.1	3.4
Ethylene chloride	9.3	3.6	2.0
Chlorobenzene	9.3	2.1	1.0
Benzene	9.1	0	1.0
Methyl ethyl ketone	7.6	4.3	2.4
Butyl acetate	7.7	2.7	0.7

For most acrylate ester polymers the three parameters are 7.5–8.0, 2.5–3.5 and 0.5–1.2. For vinyl acetate homopolymer, the values are approximately 7.6, 4.7 and 2.5. For styrene copolymers, the dispersion parameter is higher while vinyl chloride and vinylidine chloride copolymers have higher values for all three parameters.

To predict solvent systems it is necessary to consider all three parameters and if mixtures of solvent are used then it is sometimes necessary to mix hydrogen bonded and non-hydrogen bonded solvents, depending upon the nature of the polymer.

It is possible to relate solubility parameters to the polymer solvent interaction parameter χ. There are also a multitude of theories available, but none are really satisfactory and require more complicated treatments.

One area in which solubility parameters can be useful is in selecting suitable plasticisers. Similar solubility parameters of plasticiser and resin reduces problems of migration and incompatibility.

The solubility parameter theory was originally derived from the lattice theory of polymer solutions and all its inherent assumptions. The original Flory-Huggins theory assumed random distribution of contiguous segments of polymer chains. More modern polymer solution theories have diverged significantly from this original concept, and contain many correction terms for non-ideal behaviour. The concept of interaction parameter is useful for guidelines, but in practice the solubility parameter theory for useful application by the resin chemist and coating formulator leaves much to be desired. Many people do not use this approach and rely on experimentation and experience to select solvent mixtures. Others may use values of δ_s and δ_p as starting points.

Different people have differing opinions as to the usefulness of solubility parameters, but their realistic application is seriously limited.

Attempts to relate solubility parameters to viscosities of solutions have failed in practical predictions. If they had succeeded then surely they would have found everyday applications. The viscosity of solutions is proportional to the viscosity of the solvents used. Moore has shown that solubility parameters can be related to intrinsic viscosities.

As a simple and easily determined guide for solvency, the Kauri Butanol test (KB values) is probably as good as anything readily available.

DISPERSION ACRYLICS (non-Aqueous Dispersion)

These are usually dispersions of acrylic resins in aliphatic solvent, which is often a solvent for monomer but not polymer, and are termed 'N.A.D.s'. They tend to be high in molecular weight but, because they are dispersions, low in viscosity, hence coatings with high film build can be obtained. The disadvantages associated with such acrylics are complex production schedules, poor stability, lower gloss and difficult pigmentation, because the insoluble polymer particles which are dispersed and stabilised behave as if they were pigment when attempts to pigment N.A.D.s' are made. The dispersions may be either thermosetting or thermoplastic acrylics.

Early work involved dissolving material such as crêpe rubber in hydrocarbon, heating to 70–90°C, and adding premixed methyl methacrylate and peroxide initiators into the stirred mixture over a two to three hour period. After the reaction was complete the resulting acrylic was a hazy white latex type dispersion. Developments of this initial work were made through the use of:

a) Polymer stabilisers consisting of two portions, one being soluble in the diluent (aliphatic hydrocarbons), whilst the other is insoluble and anchors itself to the polymer particle. The stabilising polymer frequently is a graft copolymer either pre-formed or formed in situ. With PMMA dispersions the soluble portion can be based upon 12 hydroxy stearic acid and the insoluble anchor portion polymethyl methacrylate. The new polymer chains can be grafted onto those of the stabilising polymer.

Stabilisation is complex and depends upon stearic and electrical (charge) forces. The soluble portion of the graft copolymer extends away from the particles thereby keeping each particle away from the others. The particles are polymerised monomers which are in contact with the non-soluble portions of the stabilising polymer and can be represented as follows:

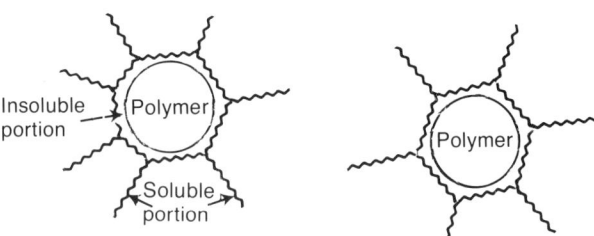

A double charge layer (zwitterion) may be formed. The greater the disparity between the components of the surface of the particle, the dispersion medium and the stabiliser, the greater the charge stabilisation.

b) An amino formaldehyde resin with a high mineral spirits tolerance is used as the aliphatic soluble portion, and the acrylic dispersions formed by adding the monomer mixture and initiators over a period of time.

The organic liquid is frequently an aliphatic hydrocarbon, e.g. white spirit, etc. Special terminology is used with N.A.D. technology and it can be confusing for conventional resin chemists. The following may illustrate this:

    The polymer (e.g. PMMA) is termed dispersion.
    The stabilising molecule (e.g. PMMA/12 hydroxy stearic acid)
    is termed dispersant.
    The solvent (e.g. heptane) is termed dispersion medium.

N.A.D.'s are being developed to overcome current and forthcoming legislation limiting solvent emission. N.A.D.'s enable high solids coatings to be applied at low application viscosities. Pigmentation is still a problem with many N.A.D.'s particularly as destabilisation with respect to time can occur due to either flocculation or charge variation.

N.A.D. is certainly a technology which will be important for certain applications in the future. But it is of minor importance today, in the ink or coating industries.

Although N.A.D.'s went some way to producing coating systems with lower levels of less toxic solvents, complexity of production and pigmentation prevented the initially, projected large useage, particularly in the automotive industry.

For further information the reader is strongly recommended to consult K. E. J. Barrett's book, 'Dispersion Polymerisation in Organic Media' (Wiley 1975, ISBN 0471054186).

High Solids Solution Acrylics

The high solids solution acrylic resins tend to be based upon modifications of the technology discussed in this chapter.

When the acrylic resins are formulated from acrylic monomers, which produce low viscosity/high solids oligomers (for example, butyl acrylate with other suitable acrylic monomers, allied to selection of initiator levels and transfer/thinning solvents), resins of acceptable viscosity at 70−80% non-volatile content may be obtained.

Cure by hexamethyl methoxy melamines or isocyanates goes a considerable way towards achieving the high solids requirements.

SOLUTION POLYMERS IN PAINTS AND INKS

Polymer solutions have many of the characteristics of conventional oleo-resinous varnishes but have two inherent disadvantages. Firstly, the high solution viscosity limits the non-volatile content of the solution and consequently the film thickness. The viscosity also depends upon molecular weight. High molecular weight polymers generally have desirable film properties, but their solutions are viscous. Thus, a compromise of the highest practical molecular weight for a given viscosity and non-volatile content solution is required. Oil based finishes oxidise to give a film insoluble in common solvents and successive coats of oleo-resinous paints can be laid down; the solvent in a new coat of a polymer solution will swell and tackify the original coat making brushing difficult. Both disadvantages can be overcome to some extent by spray application, but the process is expensive because relatively dilute solutions require large amounts of solvent to lay down films of normal thickness.

Thermoplastic acrylics are non-convertible materials which produce a coherent film by solvent evaporation under air drying or 'forced dry' conditions.

The properties of a thermoplastic acrylic resin depends almost entirely on the acrylate or methacrylate esters from which it was prepared. Many of these properties have been discussed earlier.

Thermoplastic acrylic resins may be manufactured or purchased as solutions (30−70% non-volatile content, in suitable solvents), or as bead polymers from which the end user can prepare the required solutions.

Molecular weights are of the order 30,000−130,000 for thermoplastic acrylic resins, while the thermosetting types tend to be of the order 20,000−30,000.

The viscosity of these acrylic solutions increase expotentially with molecular weight. Thus, the high molecular weight of thermoplastic acrylic resins puts severe limitations on the maximum useful solids content which can be achieved.

Thermoplastic acrylic coatings may be used for many applications. Examples of these are automotive finishes and refinishes, paper coatings, aerosols, wood lacquers, vacuum metalising lacquers, strippable coatings, inks, etc.

The development of acrylic resins, which exhibit clarity and resistance to yellowing, enabled satisfactory metallic car finishes to be formulated.

Early acrylic replacements for nitro-cellulose were deficient in adhesion and flexibility. This was later overcome by the use of special primers and plasticisers. These materials still required polishing to achieve a high standard of gloss.

The latter technology was replaced by reflow acrylics also referred to as 'bake-sand-bake' materials. Here, after the initial application of the lacquer, the coating was stoved using a short lower temperature schedule, the coating was then sanded and any imperfections repaired. The coating was then given a final stoving at higher temperatures. This caused the acrylic to reflow due to the retention of solvents and, in some cases plasticisers, giving a smooth glossy coating, free from scratches which did not require polishing. Initial reflow acrylics were subject to water spotting problems. This was eventually overcome by the use of longer ester chain methacrylates.

Clear coatings do not present any formulation difficulties, although the manufacture of acrylic solutions from beads often requires high speed dispersion and is tedious.

Pigmented coatings require good milling techniques to achieve maximum gloss and sometimes the use of predispersed pigment chips is preferred.

Acrylic resins specially formulated for ease of pigment dispersion are available.

Thermosetting acrylic resins are formulated to give films which can be cross-linked either by internal reactive groups or by reaction with an additional material.

Such resins have lower molecular weights and therefore give higher application solids than the thermoplastic acrylics. The resistance to chemicals and solvents is better with reduced softening at higher temperatures.

The film properties depend upon the monomers selected and the cross-linking functionality used.

FORMULATIONS AND METHODS FOR THE PREPARATION OF SOLUTION AND VINYL ACRYLIC RESINS

Preparation of Vinyl Acetate in Ethyl Acetate

FORMULATION

Vinyl acetate (purified)	40.00
Ethyl acetate	59.95
Azoisobutyronitrile	0.05
	100.00

PROCEDURE *(an academic polymerisation would commence here)*

1. 1.5 kg vinyl acetate is washed with 100 ml 4% sodium hydroxide solution to remove the p-methoxyphenol inhibitor.
2. After two such washings, the monomer is washed four times with 100 ml distilled water and then dried with anhydrous sodium sulphate.
3. The dried monomer is distilled using an efficient fractioning column and the first 150 ml distillate is discarded.
4. The next 750 ml distillate is collected and dried over a molecular sieve.

5. An industrial polymerisation would utilise commercial grades of monomer which would not have been treated as stages 1−4 above. The vinyl acetate and ethyl acetate are added to a polymerisation vessel equipped with stirrer and condenser.

6. Azoisobutyronitrile is added and the mixture is heated to reflux while a stream of dry nitrogen gas is passed through the mixture.

7. Determine the non-volatile content at regular intervals (e.g. every 30 minutes) to monitor the degree of conversion.

8. After three hours refluxing, the product is cooled if the non-volatile content is >38%, yielding a polymer with an intrinsic viscosity of about 1−1.3.

The solution contains about 39% polymer corresponding to a conversion of 98% of the monomer to polymer.

Both the solvent and the initiator would be carefully purified for academic work, but the details of this process are superfluous to this review. Oxygen acts as an inhibitor and nitrogen is used to exclude the oxygen in the air. In commercial operations it is not usual to remove inhibitor from the monomer before polymerisation (this is a costly and unnecessary operation). Radicals react rapidly with inhibitor which is thereby neutralised. To overcome the inhibition from the inhibitor in the monomer, the concentration of the initiator is often slightly increased. In many industrial polymerisations carried out under reflux, the reaction is not carried out under a nitrogen blanket as the solvent vapour effectively precludes the oxygen. However, if air is present in the reactor, then as the temperature is raised, contents inside the reactor will pass through compositions which are classified as explosive mixtures. Thus, it is good practice to purge a reactor with nitrogen prior to charging reactants and to maintain a steady but slight flow of nitrogen at all times.

Preparation of an Acrylic Copolymer in Methyl Ethyl Ketone

FORMULATION

Benzoyl peroxide	00.12
Methyl methacrylate	23.50
Ethyl acrylate	35.25
Methyl ethyl ketone	41.13
	100.00

PROCEDURE

1. Benzoyl peroxide is dissolved in a mixture of methyl methacrylate and ethyl acrylate.

2. 40% of the monomer mixture and methyl ethyl ketone are charged to a reactor set for reflux.

3. The reactor is heated to 90°C and ten minutes after reflux starts, the balance of the monomer mixture is added at a constant rate over a period of two and a half hours.

4. Reflux is maintained for a further two hours when a polymer with a solids content of 58% is obtained corresponding to a conversion of 98½%. If the non-volatile content is less than 58% then additional 'shots' of benzoyl peroxide are added.

The use of a delayed addition of part of the monomer is very common in industrial practice; in this instance it allows a relatively constant monomer concentration to be maintained. The addition of initiator over the same two and a half hour period gives a slight increase in the rate of radical production which will increase the rate of polymer-isation while reducing the molecular weight. The reactivity ratios of 2.0 and 0.26 (monomer one being methyl methacrylate) means that initial copolymer will be rich in M.M.A. The delayed addition of monomer mixture means that the range of copolymer distribution will be narrower than would be the case if all the monomer were added initially. The initial reflux temperature is about 85°C and will rise slightly to about 90°C towards the end of the polymerisation.

The addition of initiator to a monomer premix is commonly practised in industry but is potentially hazardous if the storage of the premix is not carefully monitored. The generation of free radicals in the premix, however caused, (possibly warming the premix or external contamination) can have catastrophic results.

If a monomer conversion curve were being constructed, then sampling non-volatile content would commence at stage three. It is good industrial practise to determine non-volatile content at the end of stage three (and preferably at least once during it) and then at hourly intervals.

Preparation of Thermoplastic Acrylic Resin in Solution

FORMULATION

Methyl methacrylate	42.0
2-ethylhexyl acrylate	9.5
Benzoyl peroxide	0.5
Xylene	24.0
Methyl isobutyl ketone	24.0
	100.0

PROCEDURE

1. Charge all solvents and 10% of the premixed monomers and 10% of initiator to the reactor with stirring.
2. Heat to 80°C − the polymerisation is exothermic − so control the temperature with cooling as necessary to below 100°C.
3. When the initial exotherm is complete add the remaining monomers and initiator as a premix over a three hour period at 90−100°C.
4. Add the final 10% of the initiator and reflux for two hours.

5. Sample for non-volatile content, if below 50% continue heating under reflux, possibly with the addition of a booster shot of peroxide until non-volatile content reaches 50%.
6. Cool and discharge through a filter press.

Preparation of Thermosetting Acrylamide Resin (Method 1)

FORMULATION

Reactor Charge	
n-butanol	26.0
Butyl glycol	8.5
Paraformaldehyde 96%	5.2
Tri-ethylamine	1.02
Acrylamide	3.9
Fumaric acid	0.85
Di-tert butyl peroxide	1.03
Monomer Tank Charge	
Styrene	33.0
Isobutyl acrylate	4.0
Dodecyl mercaptan	0.4
Di-tert butyl peroxide	1.0
AZDN	0.1
Solvesso 100	15.0
	100.00

PROCEDURE
1. Charge monomers and initiators to monomer (or header) tank and mix thoroughly.
2. Charge butanol, butyl glycol and formaldehyde to the reactor, heat to 100°C with the stirrer on to dissolve the paraformaldehyde.
3. Cool to 50°C and add the tri-ethylamine and acrylamide.
4. Add the premix from the monomer tank as quickly as possible.
5. Add the Solvesso 100 via the monomer tank.
6. Add the fumaric acid.
7. Heat slowly until exotherm begins about 85°C.
8. Turn off heat and allow exotherm to take the temperature to 120°C.

9. Reflux at 120°C for one hour and then add 80% of the di-tert butyl peroxide.
10. Reflux for a further three hours and add the remainder of the initiator.
11. Check non-volatile content every hour, when 98% plus conversion is obtained, cool and filter through press.

Preparation of Thermosetting Acrylamide Resin (Method 2)

FORMULATION

Reactor Charge	
Solvesso 100	14.5
n-butanol	5.5
Paraformaldehyde (96%)	4.0
Tri-ethylamine	0.1
Cumene hydroperoxide	0.4
Monomer Tank Charge	
n-butanol	10.0
Butyl glycol	10.0
Acrylamide	4.0
Tri-ethylamine	0.05
Styrene	10.0
Ethyl acrylate	37.0
Methacrylic acid	3.0
Di-tert butyl peroxide	0.3
AZDN	0.5
Tert dodecyl mercaptan	0.2
	100.00

PROCEDURE
1. Dissolve the acrylamide in the solvents in the monomer tank at ambient temperature. When dissolved, add the remainder of the monomer tank charge.
2. Dissolve the paraformaldehyde and tri-ethylamine in the solvents in the reactor.
3. Add 10% of the pre-mix from the monomer tank.
4. Heat to reflux (approximately 110°C).
5. Add the remainder of the pre-mixed monomers over a two hour period.

6. Add 50% of the cumene hydroperoxide and reflux for a further three hours.
7. Add remainder of the peroxide and reflux until non-volatile content is within range.
8. Cool and discharge through a filter press.

Preparation of Thermosetting Acrylamide Resin (Method 3)

FORMULATION

Reactor Charge	
n-butanol	25.0
Xylol	25.0
Acrylamide	3.75
Paraformaldehyde	4.50
Phosphoric acid	0.15
Monomer Tank Charge	
Methacrylic acid	1.6
Butyl acrylate	8.0
Styrene	23.0
Ethyl acrylate	8.0
Di-benzoyl peroxide	1.0
	100.00

PROCEDURE
1. Charge reactor with butanol, acrylamide and paraformaldehyde.
2. To dissolve − heat to 50°C with stirring.
3. Add xylol and phosphoric acid.
4. Heat charge in the reactor to 85°C and start the addition of the monomer charge, over a one hour period. Allow the temperature to rise to 100°C as the result of the exotherm and then continue heating for two further hours up to a maximum of 115°C.
5. Check non-volatile content. If 98% plus conversion, cool and filter.
6. If the non-volatile content is low add a booster shot of di-tert butyl peroxide (0.1%) and continue heating for a further hour, then recheck the non-volatile content, cool and filter as before.

Typical Characteristics

Non-volatile content	50% ± 2
Viscosity	10 – 15 poise at 25°C
Solvent blend	Xylol/butanol (1:1)

Preparation of Thermosetting Hydroxyl Acrylic Resin (Method 1)

FORMULATION

Reactor Charge	
Xylene	24.5
Ethyl glycol	24.5
Tert butyl perbenzoate	0.15
Monomer Tank Charge	
Methyl methacrylate	34.0
Ethyl acrylate	8.5
2-hydroxy ethyl methacrylate	7.5
Di-tert butyl peroxide	0.25
Benzoyl peroxide (50%)	0.60
	100.00

PROCEDURE

1. Charge all ingredients to monomer tank and thoroughly mix at ambient temperature.
2. Charge reactor with solvents and heat to 135°C while stirring.
3. Add the monomer charge from the monomer tank over a period of three hours to the solvents in the reactor at 135°C.
4. Hold at 135°C for a further two hours.
5. Add the tert butyl perbenzoate (0.15%) and hold at 135°C until non-volatile content approaches 50%.

Preparation of Thermosetting Hydroxyl Acrylic Resin (Method 2)

FORMULATION

Reactor Charge	
Toluene	25
Butyl acetate	25
Monomer Tank Charge	
Methyl methacrylate	18
Butyl acrylate	18
2-hydroxy propyl methacrylate	13
Di-benzoyl peroxide	1
	100

PROCEDURE

1. Charge reactor with toluene and butyl acetate.
2. Charge the monomer tank with stirrer left on to ensure a homogeneous mixture.
3. Heat the solvents in the reactor to 100°C and start the addition of the monomer mixture.
4. Add the monomers plus initiator over a three hour period keeping the temperature at 100°C.
5. Hold at this temperature for a further hour and check non-volatile content, continue checking non-volatile content until a value of 50% is obtained (usually in three hours from the end of monomer addition).
6. Cool and reduce to 40% non-volatile content with toluene/butyl acetate (1:1).

Typical Characteristics

Non-volatile content	40% ± 2
Viscosity	5 – 10 poise at 25°C
Solvent blend	Toluene/butyl acetate (1:1)

Preparation of Hydroxy/Carboxy Acrylic Resin

FORMULATION

Reactor Charge	
Xylene (1)	48.0
Monomer Tank Charge	
Acrylic acid	1.4
2-hydroxy propyl acrylate	7.5
Styrene	15.0
Methyl methacrylate	12.5
Butyl acrylate	12.5
Tert dodecyl mercaptan	0.1
Initiator Charge	
Di-benzoyl peroxide	1.0
Xylene (2)	2.0
	100.0

PROCEDURE
1. Charge reactor with xylene (1) and heat to 135–140°C.
2. Start the addition of the premixed monomers from the monomer tank (to be carried out at a constant rate over three hours) maintaining the temperature at 140°C.
3. Start the addition of the peroxide charge (separate feed) over the same period.
4. Hold at 140°C for one hour after the addition is complete and check the non-volatile content.
5. If non-volatile content is low – continue heating under reflux until non-volatile content is achieved. A booster shot of peroxide (0.05%) may be necessary.

Typical Characteristics

Non-volatile content	50% ± 1%
Viscosity	10 – 15 poise at 25°C
Acid value	22 mg KOH/g (on 100% non-volatile content)
Hydroxyl value	65 mg KOH/g (on 100% non-volatile content)
Solvent	Xylene

Note: There are two methods of quoting acid and hydroxyl values for vinyl and acrylic solutions. Values can be quoted either as determined on the solution, or corrected for the actual non-volatile content. Unlike resins formed by acid or hydroxyl reaction the acid and hydroxyl values should not change during processing. Although it does not matter which value is quoted provided it is clear whether it is the value of the solution or the resin, to allow for varying non-volatile contents, it is better practise to quote for 100% non-volatile content (i.e. the resin in isolation).

Preparation of Carboxyl Acrylic Resin
(with good adhesion to metal)

FORMULATION

Reactor Charge	
Toluene	45.0
Butyl alcohol	2.7
Methacrylic acid	1.5
Acrylonitrile	3.5
Ethyl acrylate	7.0
Methyl methacrylate	37.5
Initiator Charge	
Di-benzoyl peroxide	0.5
Butyl alcohol	2.3
	100.0

PROCEDURE
This process uses a method of adding peroxide shots throughout the process.
1. Charge the reactor with solvents and monomers.
2. Heat the reflux to 100°C and add the first shot of peroxide solution (0.8 total).
3. After 30 minutes add another shot of peroxide solution (0.4).
4. Add two further shots of peroxide solution (0.4) at 30 minute intervals.
5. After one hour add another shot of peroxide solution (0.4).
6. After a further hour add the final shot of peroxide.
7. Hold at temperature 100−110°C for a further hour and check the non-volatile content.
8. When non-volatile content is 49−50% cool.
9. Thin to 40% non-volatile content with toluene/butanol (1:1).

Typical Characteristics

Non-volatile content	40% ± 2
Viscosity	15 – 25 poise at 25°C
Acid value	20 mg KOH/g on (100% non-volatile content)
Solvent blend	Toluene/butyl alcohol (9:1)

FORMULATIONS FOR COATINGS AND INKS BASED ON SOLUTION POLYMERS

Some formulations taken from trade literature (and other sources) are given here as guidelines for starting point formulations, and to illustrate the principles and considerations of formulating paints and inks based upon solution vinyl polymers.

These formulations also include dissolving solid vinyl resins, however originally manufactured, in solvent and thereby forming a solution of a polymer. Thus, for example, suspension resins are being used as if they were solution ones.

In some instances this can be considered as the trivial case of a solution polymer, i.e. solid resin and solvent without polymerisation being involved.

Coating Formulations

ETCH PRIMER BASED ON POLY (Vinyl Butyral)

Industrial methylated spirits (64 O.P. I.M.S.)	31.02
Methyl ethyl ketone	20.00
To this solvent mix add with stirring:	
Poly (vinyl butyral) (p.v.b.)	9.66
Add pigment and ball-mill:	
Zinc chromate pigment	8.28
Shortly before use add a mixture of:	
n-butanol	26.21
Water	0.69
Phosphoric acid	4.14
	———
	100.00

Thin with 64 O.P. I.M.S. to a suitable viscosity for spraying

The steel or iron surface should be abraded to remove rust or scale and be degreased prior to applying the primer.

The key component in this formulation is the phosphoric acid which forms an inert barrier on iron surfaces. The polymer used must be compatible with, and unaffected by, phosphoric acid − p.v.b. meets these requirements.

Although ketones are not active solvents for butyral resins their mixtures with alcohols give lower viscosities than alcohol alone. A little water added with the phosphoric acid also enhances solubility.

Zinc chromate is often used in anti-corrosion finishes although its use is declining due to its toxicity. In the present application, it reacts slowly with phosphoric acid which requires a two pack system. The copolymer produced is a granular product which dissolves readily when added to solvent with stirring. The butanol evaporates slowly and prevents a large increase of water content during drying.

VINYL CHLORIDE STOVING ENAMEL

Mix together with stirring:	
Xylene	29.13
Vinylite VMCH	19.42
Add with stirring:	
Methyl isobutyl ketone	29.13
Incorporate on a high speed mixer:	
Tioxide R-TC90	14.56
Add with stirring:	
Epikote 1001-CX-75	2.91
Di-octyl phthalate	4.85
	100.00

The finish should be stoved at 160°C for 15 minutes

Vinylite VMCH is a terpolymer of 86% (by weight) of vinyl chloride, 13% vinyl acetate and 1% maleic acid; it is soluble in ketones, esters, nitro-compounds and chlorinated hydrocarbons. Cellosolve type esters and aromatic hydrocarbons are not solvents but can be used as diluents.

Powdered polymers are difficult to dissolve in solvents as lumps can form if the powder is added too fast; these lumps have a dry powder on the inside surrounded by a gel of swollen polymer. The technique of wetting the polymer powder in a non-solvent in which form it dissolves rapidly on adding a solvent is much used. The selection of a solvent blend involves a cost saving as xylene is cheaper than MIBK and both solvents evaporate slowly.

The stoving is used mainly to assist adhesion although it also helps to ensure complete evaporation of the solvent. The p.v.c. is 24% which ensures excellent gloss.

The addition of a stabiliser such as tri-butyl tin oxide and, possibly, a small amount of calcium carbonate might be made to improve the heat and light stability of the polymer.

ACRYLIC STOVING ENAMEL

Mill together:	
Tiona 535	26.8
Acrylic polymer solution	29.2
Xylene	7.7
n-butanol	4.0
Add with stirring:	
Acrylic polymer solution	12.4
Butylated melamine formaldehyde resin (65% non-volatile content)	17.7
n-butanol	2.2
	100.0

The finish should be stoved at 150°C for 30 minutes

Pigment volume concentration = 18%

The acrylic polymer would be a methyl methacrylate-ethyl acrylate resin containing 10–15% hydroxyethyl acrylate prepared by polymerisation in n-butanol to give a final non-volatile content of 60%.

The stoving causes cross-linking between the hydroxyl groups and the melamine resin.

Both acrylic and melamine resins have excellent colour retention while the cross-linking hardens and toughens the film.

THERMOPLASTIC ACRYLIC CLEAR LACQUER

Acrylic copolymer (50% non-volatile content)	65
Butyl benzyl phthalate	3
Toluene	16
MEK	16
	100

This acrylic is suitable for use as a protective lacquer for vacuum deposited metals.

VINYL FINISH FOR BASE STEEL

Vinylite VMCH	17
Tri-cresyl phosphate	4
Rutile titanium di-oxide	13
Xylene	33
Methyl ethyl ketone	33
	100

Pigment volume concentration = 17%   *Volume solids = 20%*

This formulation is also suitable for phosphated steel, non-ferrous metals and concrete.

VINYL ANTI-FOULING PAINT

Xylene	20.78
Add with stirring:	
Vinylite VAGH	10.39
Increase the stirring rate and add:	
Methyl isobutyl ketone	23.38
Reduce stirrer speed and add sequentially:	
n-butanol	1.30
Rosin	5.19
Tri-butyl tin oxide	10.39
Tri-cresyl phosphate	5.19
Ball mill for 20 hours with:	
Copper suboxide	23.38
	100.00

Vinylite VAGH is a copolymer containing 91% vinyl chloride, 3% vinyl acetate and 6% vinyl alcohol as its constituent monomers. The principal function of the binder in an antifouling paint is to permit a slow diffusion of toxins from its surface.

The tri-butyl tin oxide should be compatible with the polymer to permit slow migration to the surface of the film to replenish material which has dissolved in the sea water.

Some water sensitivity in the film will allow very slow dissolution of the copper suboxide into the surrounding water and this is provided by the hydroxyl groups in the polymer. Note that this coating is not intended to be anti-corrosive and that its adhesion to the undercoat need be only moderate.

Tri-butyl tin oxide is extremely harmful to marine life, particularly shellfish, at relatively low levels. The use of Tri-butyl tin oxide in anti-fouling paints is becoming more restricted particularly for craft moored in estuaries. It is likely that these restrictions will increase and it is anticipated that tin based anti-fouling paint will be restricted to hulls of ocean-going craft only. An example of the effects is that oysters which cycle between being male and female grow a penis during the female cycle, thereby inhibiting their breeding.

WHITE STOVING ENAMEL
FOR DOMESTIC APPLIANCES

Thermosetting acrylamide resin*	52.0
Epikote 1001 (50% in ethyl/glycol acetate)	6.0
TiO_2	35.0
Solvesso 100	7.0
Silicone oil	trace
	100.0

*60% solution in xylol/butanol (1:1)
Stove for 30 minutes at 150°C

METALLIC BASE COAT FOR AUTOMOTIVE USE

Thermosetting hydroxy acrylic resin*	24.0
CAB (50% in xylene/butyl acetate)	35.0
Aluminium paste (65%)	4.0
Melamine formaldehyde resin (72% in butanol)	2.5
n-butyl acetate	4.0
Thinning solvents	30.5
	100.0

*50% solution in xylene/butanol (3:1)
Stove for 30 minutes at 130°C

CLEAR TOP COAT FOR AUTOMOTIVE USE
(Refinishing)

Thermosetting hydroxy acrylic resin*	75.0
Ethyl glycol acetate	11.5
Xylene	13.0
Tinuvin 328 (Ciba-Geigy)	0.5
	100.0

*60% resin in xylene/ethyl glycol acetate

Cure with aliphatic polyisocyanate   14 p.b.w.
Desmodur N ex Bayer Ltd.

WHITE COATING FOR WOOD

Thermosetting hydroxy acrylic resin*	40.0
TiO_2	25.0
Polypropylene wax	0.5
Thinning solvents (xylene, toluene, ethyl acetate, MIBK and MEK)	34.5
	100.0

*50% resin in xylol/MIBK

Cure with an aliphatic polyisocyanate   6.5 p.b.w.
Desmodur N in 12 p.b.w. thinners.

DEEP-DRAWING ENAMEL FOR
TIN PLATE AND ALUMINIUM

Thermosetting acrylamide resin (60%)	49.0
TiO_2	33.0
Epikote 1001 (50%)	11.0
Modaflow (30% Solvesso 150)	0.5
Ethyl di-glycol acetate	2.5
Solvent (Solvesso/butanol (1:1)	4.0
	100.0

Stoving schedule 10 minutes at 190°C

PROTECTIVE ACRYLIC COATINGS
FOR COPPER METALS

Acryloid B44S (40%)*	74.40
Toluene	19.72
Ethyl alcohol	5.00
Benzotriazole (chelating agent)	0.44
Paraplex G60 (levelling agent)	0.44
	100.00

*Acryloid B44 is a methyl methacrylate copolymer, available as bead polymer or solution from Rohm & Haas Ltd.

Reduce to 11–13% non-volatile content with toluene, spray apply to yield a final film thickness of 1.0 mil.

COATING FOR ABS PLASTICS

Acryloid B72 (100%)*	6.0
CAB 553 – 0.4**	3.0
CAB 381 – 0.5**	6.0
Pigments	5.0
Toluene	40.0
Ethyl glycol acetate	10.0
Ethanol	8.0
Ethyl acetate	8.0
Isobutyl acetate	14.0
	100.0

*Acryloid B72 is ethyl methacrylate copolymer (100%) from Rohm & Haas Ltd.

**CAB 553 and 381 are grades of cellulose acetate butyrate from Eastman Chemicals Ltd.

WOOD FINISHING LACQUER

Acryloid B72 (50%)*	50
$\frac{1}{2}$S RS nitro-cellulose	35
Paraplex G50	15
	100

*Acryloid B72 is ethyl methacrylate copolymer,
50% in toluene from Rohm & Haas Ltd.

This formulation may be brush applied or reduced with
lacquer solvents for spray application.

PRINTING INK APPLICATIONS

Emulsions and especially polymer emulsions are little used in printing inks because drying can cause an irreversible build-up on the rollers. Flexographic and letterpress inks provide the main use for solutions of vinyl polymers although some polymers find use in the gravure process.

Gravure Inks

The gravure process requires a low viscosity ink while pigment dispersion requires high resin contents. Low molecular weight polymers are mandatory for these applications and even then the viscosity is higher than for similar concentrations of rosin based inks. Vinyl polymers are only used when they provide some special property.

Solvent evaporation is extremely important in gravure inks, while evaporation in the ink duct is not a major factor if the duct is covered. On films or foils where solvent penetration into the substrate is slow, it is particularly important that the ink should not flow reducing the sharpness of the print pattern. The thinness of the applied coating is the main factor in preventing the spread of ink, but the rapid loss of some of the solvent would be an additional influence. During this initial period after application, the printed pattern must flow to give a print of even thickness by the action of surface tension forces on the irregular surface transferred from the cylinder cells. Drying of the ink is necessary to prevent smudging of the print during processing and reeling, and this can be achieved by blowing hot air across the print surface. Other methods of forced drying such as passing over heated rollers or passing through an oven are liable to damage thermo-plastic film.

Printers like to run their presses as fast as possible for economic reasons and solvent release to achieve a tack free film to enable reeling or stocking tends to dictate the print speed. Different solvents or solvent mixtures are used but there is a trend away from some of the more volatile and harmful solvents which limits press speed.

Solvent recovery is often achieved by absorbing the solvent from the ink on charcoal, thus reducing both fire and pollution risks. On heating the charcoal the solvents can be distilled and condensed to give a solvent blend suitable for thinning later batches of inks. The above formulation might need thinning from 60% to as much as 75% total solvent depending on the type of press used.

A useful rule of thumb in gravure inks is that proportions of pigment and resin should be approximately equal. With the relatively low total solids and the very thin films which are applied, pigments of high tinctorial power are required, and the use of extenders is prescribed. The choice of pigments is essentially limited to organic toners which do not bleed or crystallise in the solvents used.

RED GRAVURE INK FOR PVC FILM

Toluene	24.0
Stir and add slowly:	
Vinylite VYHH	10.0
Vinylite VYLF	10.0
Increase stirrer speed and add:	
Methyl ethyl ketone	14.0
Methyl isobutyl ketone	19.0
Reduce stirrer speed and add:	
Di-octyl phthalate	4.0
Lake Red C	19.0
Incorporate pigment in a high speed mixer	100.0

Both vinylite resins are copolymers (87% vinyl chloride and 13% vinyl acetate). A mixture of polymers of medium and low molecular weight gives a compromise between film strength and solution viscosity. The polymer has good adhesion to plasticised p.v.c., but plasticiser migration from the film into the ink can soften the print and cause blocking in the reel.

YELLOW GRAVURE INK FOR POLYESTER FILM

Polymer solution (60% non-volatile content)	25.0
Isopropanol	10.0
Rutile titanium di-oxide	4.5
Hansa Yellow G	10.5
Incorporate on a high speed mixture then add with stirring:	
Isopropanol	50.0
	100.0

The polymer solution is prepared by the solution polymerisation of 85% vinyl acetate and 15% methyl hydrogen maleate in isopropanol.

The isopropanol gives a low molecular weight copolymer because of its relatively high chain transfer constant. A vinyl acetate homopolymer tends to be tacky and the acid maleate increases the Tg. The acid groups promote adhesion to the polyester.

Inks prepared from organic pigments have low opacity and thus give a transparent effect on films; for yellows some opacity may be produced by adding white pigments.

FLUORESCENT GRAVURE INK

Acryloid NAD – 10*	30.0
Fluorescent pigment	45.0
Toner	1.0
White spirit	13.0
Toluene	1.0
Thickener (100%)	1.0
Textile spirits/toluene (9:1)	9.0
	100.0

*Acryloid NAD – 10 is a non-aqueous dispersion, 40% non-volatile content in white spirit from Rohm & Haas Ltd.

VINYL AND ACRYLIC
AQUEOUS POLYMERISATION

The low cost of vinyl polymers coupled with their ease of production and the very wide range of chemical types available with a corresponding range of physical properties, all contribute to their widespread use in surface coatings. Most of the vinyl polymer usage in coatings involve water based finishes which are both cheaper and produce less pollution than solvent based products. The ability to produce emulsions directly from monomer by a polymerisation process rather than by post-polymerisation emulsification has introduced new technological techniques to the paint industry. In the expanding field of alkali soluble coatings, vinyl polymers have considerable advantages over alkyd solutions.

The polymerisation process produces higher molecular weights than are obtained by condensation polymerisation. Variations are obtained by using different monomers and a wide range of polar and functional groups can be obtained. The principle monomers used in the production of coating polymers are:

Ethylene	$CH_2 = CH_2$
Vinyl chloride	$CH_2 = CHCl$
Vinyl acetate	$CH_2 = CHOCOCH_3$
Acrylonitrile	$CH_2 = CHCN$
Acrylic acid	$CH_2 = CHCOOH$ and its methyl, ethyl, butyl and 2-ethylhexyl esters

Methacrylic acid $\quad CH_2 = CCOOH$ and its methyl ester
$$\underset{CH_3}{|}$$

Styrene (vinyl benzene) $\quad CH_2 = CHPh$ (where Ph denotes the nucleus of a benzene molecule)

Vinylidine chloride $\quad CH_2 = CCl_2$

Maleic anhydride

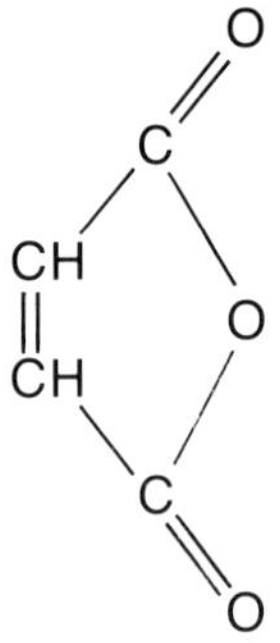

and its half and di-esters

Water Soluble Vinyl and Acrylic Polymers

The difference between a true aqueous solution of a high molecular weight polymer and an emulsion is clear. In the emulsion, a miscelle consisting of a number of polymer molecules is solubilised by a surface covering of hydrophillic molecules, but in true solution a single polymer molecule is dispersed on its own and is completely surrounded by water. However, if we consider the continuous reduction in particle size from an emulsion miscelle to a single molecule, there is a zone of colloidal dispersion where the miscelle size is so small that it is invisible to the eye and the dispersion is clear.

Very few polymers form true solutions in water, thus this section will include all clear polymer dispersions which remain clear when diluted with water to very low concentrations. Although there is a continuous progression of properties from water soluble resins through colloidal dispersions to aqueous emulsions, it is possible to tabulate the typical properties of each.

Typical Property	Water Soluble	Colloidal Dispersion	Aqueous Emulsion
Molecular weight	20,000 – 50,000	20,000 – 200,000	100,000 – 1,000,000
Viscosity	Dependent on molecular weight	Dependent on molecular weight and on pH	Low, and independent of molecular weight
Non-volatile content at application Viscosity	Lowest	Intermediate	Highest
Rheological Properties	Newtonian	Nearly Newtonian	Pseudoplastic
Chemical resistance (not cross-linked)	Fair	Good	Excellent
Flexibility	Fair	Fair – Good	Excellent
Durability	Good	Good	Excellent
Gloss	Good	Good	Poor

An organic polymer is only soluble in water if it contains polar groups. Also, a low molecular weight helps to make a polymer more soluble. Therefore, thermoplastic water-soluble polymers are of limited use in coatings, because the hydrophillic groups would make the film water-sensitive and the low molecular weight would detract from the durability and flexibility of the film. These disadvantages are overcome in thermosetting polymers by cross-linking the film.

Polyvinyl alcohol is water soluble and has film forming properties. Unfortunately, the films are water sensitive and therefore of limited use. The main use of P.V.A. in surface coatings is as a thickener or as a stabiliser in emulsions. However, a number of coating systems have been developed in which some of the hydroxyl groups are cross-linked to form insoluble films. The cross-linking agents include light-sensitive diazonium salts, metal salts, urea-formaldehyde resins, phenol-formaldehyde resins, organo-titanates and other reagents which will link hydroxy groups at ambient or elevated temperatures.

222

Example 1

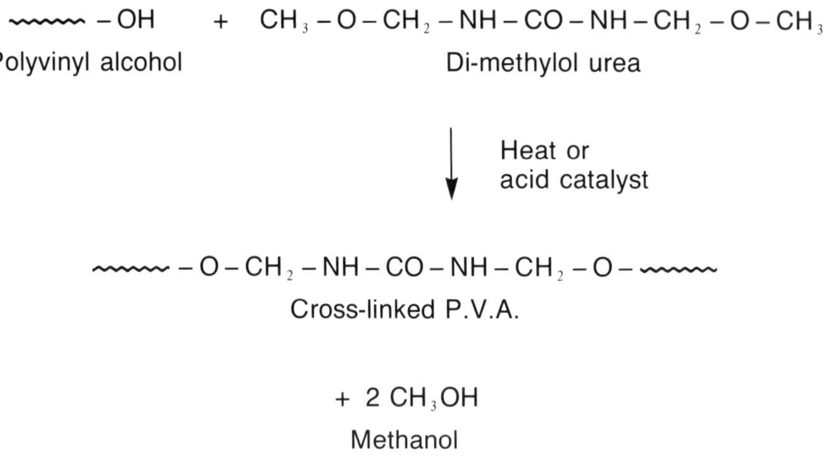

〰〰〰 – OH + $CH_3 – O – CH_2 – NH – CO – NH – CH_2 – O – CH_3$

Polyvinyl alcohol   Di-methylol urea

↓ Heat or
  acid catalyst

〰〰〰 $– O – CH_2 – NH – CO – NH – CH_2 – O –$ 〰〰〰

Cross-linked P.V.A.

$+ \ 2 \ CH_3OH$

Methanol

Example 2

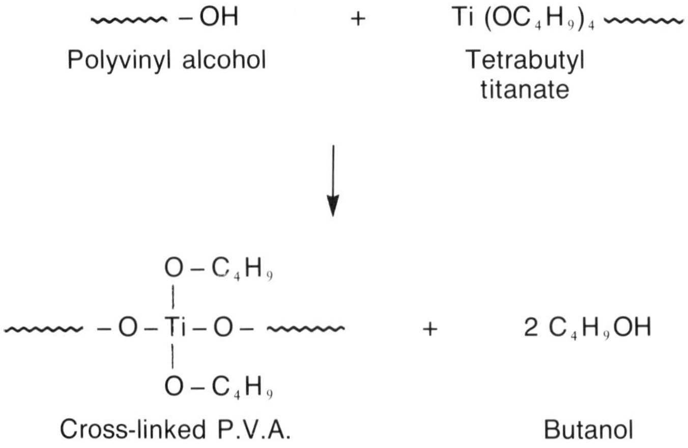

〰〰〰 $– OH$ + $Ti \ (OC_4H_9)_4$ 〰〰〰

Polyvinyl alcohol   Tetrabutyl
                                      titanate

↓

$$\begin{array}{c} O – C_4H_9 \\ | \\ 〰〰〰 – O – Ti – O – 〰〰〰 \\ | \\ O – C_4H_9 \end{array}$$ + $2 \ C_4H_9OH$

Cross-linked P.V.A.   Butanol

An alternative way of making P.V.A. films insoluble in water is to remove the water sensitive hydroxyl sites by converting them to acetals with aldehydes, or complexing them with basic acid.

$$\text{\large\ensuremath{\sim\sim}} - CH - CH_2 - CH - CH_2 - \sim\sim \qquad + \qquad RCHO$$
$$\qquad\ |\qquad\qquad\ |$$
$$\qquad OH \qquad\quad OH$$

Polyvinyl alcohol   Aldehyde

↓

$$\text{\large\ensuremath{\sim\sim}} - CH - CH_2 - CH - CH_2 - \sim\sim$$
$$\qquad\ \backslash \qquad\qquad\ /$$
$$\qquad O - CH - O$$
$$\qquad\qquad |$$
$$\qquad\qquad R$$

Acetal group

Polyvinyl alcohol is made by hydrolysis of the acetate groups on polyvinyl acetate, so the degree of hydrolysis can be used to control the water solubility properties. The types used commercially to stabilise polymer emulsions usually have 10–15% of the acetate unhydrolysed.

Polyvinyl pyrrolidone is another water soluble polymer which gives films of limited use because of their water sensitivity. Also, vinyl pyrrolidone is too expensive for most applications.

Water soluble vinyl or acrylic polymers can be produced by incorporating carboxyl groups along the chain and by limiting the molecular weight by means of a chain transfer agent. These polymers may be made either by the emulsion process or the solvent solution process and subsequently dissolved in water. For solubilisation after neutralisation a minimum acid value of 50–60 mgkOH/g is required, as a general rule.

It is also possible, but much less common, to produce a water soluble acrylic from a water insoluble acrylic by converting hydrophobic groups to hydrophilic pendent groups. For example, esters, amides or nitrile groups can be hydrolysed or unsaturated groups oxidised.

Carboxyl containing acrylic polymers are often stabilised in aqueous solutions by adding ammonium or amines to form salts. These amines or ammonia can be largely driven off after film formation as described in the section on water soluble alkyds. However, even after the loss of these nitrogen compounds, the films remain water sensitive and some form of cross-linking is necessary for water resistance.

The types of cross-linking agents previously described can be used. The most useful are the N-methyol group containing resins such as the urea or melamine formaldehyde resins and the phenol formaldehyde resins which react with hydroxyl or carboxyl groups on the polymer. The reactions with polyvalent salts or with the hydroxyl groups on polyvinyl alcohol have also been used.

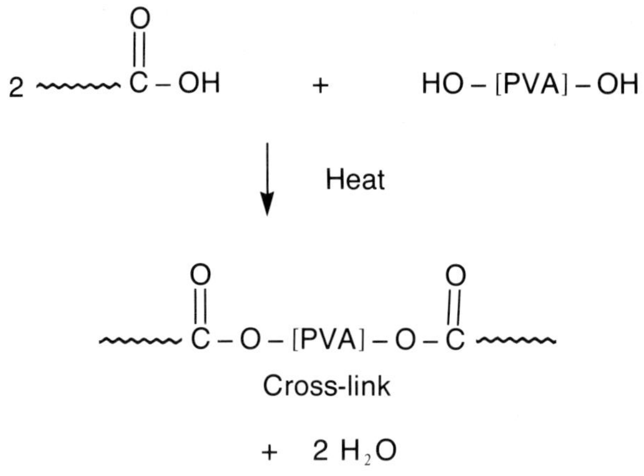

$$2 \sim\sim\sim \overset{\overset{\displaystyle O}{\|}}{C} - OH \quad + \quad HO - [PVA] - OH$$

↓ Heat

$$\sim\sim\sim \overset{\overset{\displaystyle O}{\|}}{C} - O - [PVA] - O - \overset{\overset{\displaystyle O}{\|}}{C} \sim\sim\sim$$

Cross-link

$$+ \quad 2\,H_2O$$

Aqueous Solution Polymerisation

For a few resins which are water soluble and their components are also water soluble, it is possible to prepare them in water.

Chemical modification of vinyl polymers is not common. The hydrolysis of poly(vinyl acetate) to the alcohol is used to provide a valuable colloid. The poly(vinyl alcohol) can be reacted with aldehydes to produce cyclic acetals which have several uses.

LABORATORY SCALE PREPARATION OF POLY(VINYL BUTYRAL)

Poly(vinyl alcohol)	5.5
Water	82.0
Methanol	3.6
Sulphuric acid	0.2
Butyraldehyde	8.7
Sodium carbonate solution	to neutralisation
	100.0

PROCEDURE

1. 100 g poly(vinyl alcohol) is dissolved in 800 g water.
2. 65 g methanol and 0.3 g sulphuric acid are added.
3. 60 g butyraldehyde are added with gentle stirring.
4. 250 g of this solution are placed in a two litre flask.

5. 100 g butyraldehyde are added with vigorous stirring followed by the gradual addition of the remainder of the polymer solution over 30 minutes. During this addition the temperature will rise to about 70−80°C.
6. 700 g water containing 3 g sulphuric acid is heated to 75°C and added over 15 minutes to the hot reaction mixture. Vigorous stirring is essential during this addition.
7. The mixture is stirred more slowly for a further hour and allowed to cool before filtering off the product.
8. The precipitate is washed with dilute sodium carbonate solution to remove the last traces of sulphuric acid and dried. About 200 g product are obtained.

The reaction is:

$$\sim\!\!\sim\!\!\sim CH - CH_2 - CH - CH_2 \sim\!\!\sim\!\!\sim \ + Pr.CHO$$

(with OH groups on the first and third CH carbons)

$$\downarrow$$

(cyclic acetal structure with Pr—C at top, O atoms, CH and CH—CH$_2$, and CH$_2$ bridge)

Polyvinyl acetals are characterised by molecular weight and by composition, which is normally expressed as the acetate content remaining after the original hydrolysis plus the aldehyde content. The only other chemical modification of commercial importance is the production of carbon fibres from polyacrylonitrile.

LABORATORY SCALE AQUEOUS POLYMERISATION OF N-VINYL PYRROLIDONE

N-vinyl pyrrolidone	19.0
Water (deionised)	80.7
Potassium persulphate	0.2
Sodium acetate	0.1
	100.0

PROCEDURE

1. 200 g N-vinyl pyrrolidone and 800 g deionised water are added to a two litre flask fitted with a stirrer, reflux condenser and nitrogen inlet tube. 1 g potassium persulphate is added and the stirrer and nitrogen flow are started.
2. The flask is heated in a water bath maintained at 80°C. After 30 minutes at 80°C, a solution of 1 g sodium acetate and 1 g potassium persulphate in 50 g water is added through the condenser over two hours, and the reaction is continued for a further two and a half hours.
3. The reflux condenser is moved to allow distillation and the water evaporated under reduced pressure. The product is a clear resinous mass.

Even when the reaction is not under reflux, it is normal to use a condenser when nitrogen is being used, as it becomes saturated with solvent vapour at the reaction temperature. The condenser minimises solvent loss. The sodium acetate is used as a buffer because some of the persulphate can decompose to give sulphuric acid and oxygen. Notice the relatively long reaction time which is typical of aqueous polymerisations.

Another approach to preparing aqueous solutions and one which is typical of the industrial approach is to prepare the resin as an addition solution polymer, and then remove or partially replace the organic solvent. Either distillation of solvent prior to neutralisation of the resin and subsequent addition of water or neutralisation, and addition of water prior to distillation of organic solvent possibly as an azeotrope. The major disadvantage of the former approach is the high viscosities of the remaining resin solution prior to neutralisation. As a generalisation some organic solvent always remains in the final resin solution and normally water miscible solvents are used.

LABORATORY SCALE PREPARATION OF AN ACRYLIC COPOLYMER CONTAINING FUNCTIONAL GROUPS

Acrylamide	1.7
Itaconic acid	3.4
Isopropanol	33.9
Ethyl acrylate	28.1
Benzoyl peroxide	0.3
Deionised water	28.1
Ammonia (27%)	4.5
	100.0

30 g acrylamide and 60 g itaconic acid are dissolved in 300 g isopropanol; after filtering, 500 g ethyl acrylate are added and 6 g benzoyl peroxide are dissolved in the mixture.

PROCEDURE

1. 300 g isopropanol are added to a two litre flask fitted with a stirrer, reflux condenser and tap funnel.
2. The flask is heated until the isopropanol starts to reflux when the monomer mixture is added at a constant rate over two hours; reflux is continued for a further 30 minutes giving 97% conversion.
3. The reflux condenser is moved and set for distillation.
4. 450 g isopropanol are distilled off when 200 g deionised water are added. Distillation is continued to remove the balance of the isopropanol as an azeotrope containing 92% isopropanol.
5. 80 g of a 27% solution of ammonia and 300 g water are added with stirring giving an aqueous solution of 49−50% non-volatile content.

The larger proportion of initiator as compared with some of the examples, tends to give more rapid reaction and a lower molecular weight. Note that the method of monomer addition allows a large proportion of the copolymer to have a random sequence of monomer units with the overall composition of the added mixture. Isopropanol is often used to produce copolymers of low molecular weight owing to its high chain transfer activity, although quantitative values are not available. Another interesting point in this example is the distillation of the (expensive) solvent and the eventual production of a solution in water which is very cheap. The distillation would not be possible for a higher molecular weight polymer and it does require a high energy input.

PREPARATION OF AN AQUEOUS POLYMER AND AN AQUEOUS ANTI-CORROSIVE SPRAY PAINT UTILISING IT

Polymer solution (70% non-volatile content)	24.65
Add with stirring:	
Di-methylaminoethanol	1.40
Water	44.37
Hexamethoxymethyl melamine, 50%	14.79
Water	14.79
The above is the binder solution	100.00

PREPARATION OF AN AQUEOUS
ANTI-CORROSIVE SPRAY PAINT BASED
ON PREVIOUS AQUEOUS POLYMER (cont.)

Ball mill for 24 hours*	
Binder solution (from above)	18.67
Kronos RN-56	16.81
To the milled paste add: with stirring:	
Binder solution	31.58
Tegopren	0.037
Ammonium p-toluenesulphonate	0.075
Dilute to 30% non-volatile content with:	
Water (de-ionised)	32.828
	————
	100.000

PROCEDURE

1. A copolymer containing 350 g Veova 10, 46.2 g acrylic acid, 72.1 g hydroxyethyl acrylate, 120 g methyl methacrylate and 112 g styrene is prepared by solution polymerisation in 300 g of butyl cellosolve. 57 g di-methylaminoethanol are stirred in and 1800 g water added slowly with stirring to give a solution at 25% solids. 600 g of a 50% solution of hexamethoxymethyl melamine and 600 g water are added. This solution is the binder solution.

2. 1350 g Kronos RN-56 and 1500 g binder solution are ball-milled for 24 hours when the balance of the binder solution is added.

3. 3 g Tegopren and 6 g ammonium p-toluenesulphonate are added and the paint diluted to 30% solids with de-ionised water.

In this example a solution polymer is prepared using standard techniques. The polymer is then solubilised by neutralising the acrylic acid with D.M.A.E. Although the paint is claimed to be aqueous there is a proportion of solvent (butyl cellosolve) present. This technique of preparing aqueous vinyl polymers is quite common.

The use of di-methylaminoethanol to neutralise the acid groups of the polymer is preferred to ammonia in stoving finishes. Ammonium salts of carboxylic acids can give amides on heating and they also react with formaldehyde, reducing the efficacy of the cross-linking resin. On standing, the pH of ammonia solutions tends to drop with a consequent rise in viscosity, whereas D.M.A.E. solutions have good viscosity stability on ageing. The sulphonic acid salt is added as a latent catalyst which reverts to the acid on heating. The Tegopren is a flow control additive which prevents pin holing.

The polymer composition raises several interesting points. The hydroxyethyl acrylate reacts more readily with melamine resins than carboxylic acids. The acrylic acid is added to give water solubility when neutralised. Note that in the absence of the hydroxy monomer, more acrylic acid would be required to solubilise the polymer. The Veova has a softening effect on the copolymer and also contributes water resistance to the film. This paint has anti-corrosion properties similar to those of conventional paints containing chromate or other pigments which inhibit rusting.

For spraying use a larger nozzle than usual, e.g. 3.5 mm, because of the relatively high viscosity. The sprayed film is stoved at 160°C for 30 minutes.

SUSPENSION POLYMERISATION

If an acrylic monomer or monomer mixture is insoluble in water, bulk polymerisation can be carried out in the suspended droplets, using the surrounding water as a heat transfer medium. Acrylic monomer is obtained in the form of small beads, prills, pearls or granules.

A monomer phase consisting of a mixture of monomer, monomer soluble initiator and chain transfer agent is suspended in the form of droplets in a non-solvent. The use of organic non-solvents has been discussed under N.A.D.'s. Commercial processes utilise water as the non-solvent. The water or aqueous phase contains selected dispersing agents such as polyvinyl alcohol or gelatine, and various salts such as di-ammonium hydrogen phosphate and sodium carbonate which act to increase interfacial tension between the monomer and aqueous phases, and so prevent the monomer droplets coalescing.

Polymerisation occurs within each monomer droplet and continues until all the monomer is consumed. At which point the original droplet is in the form of a polymer bead. The overall effect could be considered to be a bulk polymerisation (carried out in a multitude of discrete individual droplets) with each droplet being surrounded by water which acts as a heat sink to dissipate the heat of reaction. The water can also aid agitation of the reactants. The agitation generated by the stirrer is of major importance not only in keeping the polymer beads in suspension in the aqueous phase, but the shear induced controls the bead size. A fast stirring rate results in the monomer drops being shattered into small droplets and ultimately results in very small polymer beads, whilst a slow stirring rate results in the monomer droplets coalescing to form large beads. Once the reaction is complete the suspension of polymer beads is run from the reactor into holding tanks, where in the absence of agitation the polymer beads 'settle out' of the aqueous phase. The beads are washed free of contaminating salts and suspending agents, and are dried in specially designed plants.

The suspension polymerisation technique places certain restrictions on the choice of initiator. Since water is present, ionic initiators are unsuitable for use in suspension polymerisation. Therefore the choice is restricted to free radical initiators. The choice is limited further by the necessity of the initiator to be monomer soluble and water insoluble.

The polymerisation temperatures normally encountered are in the range 60−85°C. This relatively low reaction temperature necessitates the use of initiators with short half-life times. Perpivalates, perdicarbonates, peroctoates and azobis isobutyryl nitrile are a few of the more commonly encountered suspension polymerisation initiators. When azobis isobutyryl nitrile (AZBN) is employed there is the possibility of the nitrogen eliminated during the decomposition of the AZBN giving rise to the formation of hollow beads. The hollow beads themselves are not deleterious in terms of their application, however, the reduction in density experienced with hollow beads means that they are liable to float in the 'settling-out' tanks leading to practical problems during the washing and filtering stages of manufacture.

Suspension polymerised resins are not widely used in surface coating applications, although there is a large specialised use area in printing ink formulations which are intolerant to the presence of the types of solvent normally encountered with solution processed polymers, but all traces of stabilisers and residual water need to be removed.

ADVANTAGES
   i) The molecular weight distribution is very narrow.
  ii) Both high or low molecular weight polymers can be produced as required.
 iii) Polymers can be prepared as a solid, thus allowing the surface coatings formulator a free choice of solvents when compounding the coating.

DISADVANTAGES
   i) The polymer has to be subjected to a lengthy washing and drying procedure to remove the salts and surfactants present from the aqueous phase.
  ii) The ratio of monomer phase : aqueous phase, normally employed is of the order ca 35:65. This fact coupled with the product losses, during the washing and drying procedures, result in typical product yields of ca 30% of the original reactor charge.
 iii) The polymerisation process requires rigid control of temperature and stirring rate, in order to prevent coalescence of the beads during manufacture.

Plant

This normally consists of a stainless steel reactor, very similar to that used for solution polymerisation, fitted with an outer jacket through which steam or water may be passed. In addition to the ancillary plant, mentioned above, provision of an inert gas inlet for blanketing with a nitrogen atmosphere is necessary. The essential feature of this plant is the provision of an efficient high speed stirrer (150−300 rpm) possibly with baffles fitted into the reactor walls. In specialised plant, agitators can have 'wiper blades' fitted to clean the reactor walls. It is necessary to have a variable speed agitator. It is necessary to ensure that the discharge system can handle the resulting slurry of bead polymer and water additives.

Process

The reactor is charged with the required amount of water (usually deionised) and the suspending agent added. The aqueous phase is heated to about 80°C under an inert gas atmosphere. The monomer mixture and initiator (20−40% of the total charge depending upon the monomer and the degree of exothermic activity of the reaction), is added slowly from a monomer tank with the stirrer running at maximum (or pre-determined) speed. After a period of time, usually less than an hour, an exotherm occurs. This phenomenum may be slight or considerable, depending upon the process conditions. The reaction usually proceeds rapidly and is almost complete once the exothermic phase of the process is over.

Initially the monomer phase is lighter than water and surface agitation is necessary. Particles of the slurry are sticky and will agglomerate. After the 'exotherm' the polymer phase is heavier than water and can settle out if the stirring is inadequate.

The suspension is cooled and the beads recovered either by filtering through a stainless steel screen or by centrifuging. After thorough washing, the beads are dried at 80−120°C in an oven, various heated driers or vacuum driers.

In most cases a high degree of conversion is obtained and bead like polymers result, usually about 100−1000 μm in diameter.

Since water is the continuous phase, viscosity changes very little with conversion, so that the heat transfer to the reactor cooling system is very efficient. Initiators which are insoluble in water (oil soluble initiators) are used for suspension polymerisation. Emulsion polymerisation requires water soluble initiators. Initiators which are soluble in the monomer phase provide a much faster rate of polymerisation than the water soluble ones used for emulsion polymerisation.

The suspensions are produced by continuous efficient agitation of the monomer in the solvent phase (water). The behaviour inside the droplets is very similar to that which occurs in bulk polymerisation but, because the droplets are only 10 to 1000 μm in diameter, very rapid rates of reaction can be tolerated without the monomer boiling.

The particle size and size distribution of the polymer is affected by the stirring rate, suspending agent, impurities and interfacial tension. Electrostatic charges are induced on the suspended monomer/polymer particles which retard coalescence. Workers have found that the particle diameter varies inversely with the impeller tip speed. The polymer remains dissolved in the monomer droplets. As the reaction proceeds, these droplets increase in viscosity and stickiness.

As the polymerisation proceeds the droplets pass from the liquid to solid state passing through an intermediate sticky stage. At about 15−30% monomer conversion the sticky phase is reached. This persists up to about 70−85% conversion when the solid phase is reached. The protective colloid (suspending agent) is necessary to prevent the particles coalescing, particularly during the initial stages, even so this does not prevent the particles from rising ('creaming') together if agitation is stopped. 'Scale up' from laboratory to production of suspension polymers may be fraught with problems. It is necessary for the chemist to understand the factors affecting the size of beads and the stability of the reaction mixture during polymerisation at laboratory scale, in order to predict events at commercial scale.

Formulations for Suspension Polymers

EXAMPLE 1   **PREPARATION OF A SUSPENSION POLYMER OF METHYL METHACRYLATE**

Aqueous phase	
De-ionised water	79.7
Suspending agent (polyvinyl alcohol or sodium polyacrylate)	0.2
Oil Phase	
Methyl methacrylate	19.9
Di-benzoyl peroxide	0.2
	100.0

PROCEDURE
1. Charge water and polyvinyl alcohol, heat to 80°C with vigorous stirring and under inert gas.
2. Add the methyl methacrylate and the initiator premixed.
3. After about one hour a slight exotherm will occur, this can be detected by 10−15°C increase in temperature.
4. Continue stirring for four hours and then apply cooling.
5. Filter off methacrylate beads (200−900μ) wash well with water and dry in an oven at 70°C.

EXAMPLE 2 **PREPARATION OF A POLYVINYL CHLORIDE SUSPENSION POLYMER**

Vinyl chloride	35.00
Polyvinyl alcohol	0.40
Di-benzoyl peroxide	0.50
Water	64.91
	100.00

PROCEDURE
As for Example 1, but polymerise for 10 hours at 60°C in an autoclave.

EXAMPLE 3 **PREPARATION OF VINYL CHLORIDE/VINYL ACETATE SUSPENSION COPOLYMER**

Vinyl chloride	35.0
Vinyl acetate	7.0
Vinyl alcohol/acetate, 80/20 copolymer	0.1
Lauroyl peroxide	0.2
Water	57.6
Sodium lauryl sulphate	0.1
	100.0

PROCEDURE
As for Example 1, but polymerise under autogenous pressure for 12 hours at 50°C.

Formulations Utilising Suspension Polymers

Suspension polymers for surface coatings are insoluble in water and require organic solvents to form a liquid system. Therefore, formulations showing some of the applications of suspension polymers have been given in the solution polymer formulary.

EMULSIFIED POLYMERS

In addition to forming emulsion polymers in situ it is possible to emulsify some types of polymers. This depends upon the polarity, presence of surfactants and other factors. The scope of this is outside the content of the present chapter and various types of emulsified resins are discussed in the Aqueous Polymer chapter, Volume III.

For our purposes emulsion polymerisation and emulsion polymers will be restricted to vinyl and acrylic polymerisation.

EMULSION POLYMERISATION

Resin-water emulsions can be produced by post-reaction emulsification of a condensation or addition polymer by the use of surfactants or emulsifiers to form an 'oil-in-water' emulsion of the polymer in the aqueous phase. This is simply a method of obtaining a polymer in an aqueous medium. This type of emulsion is different to a true emulsion polymer made by the process of emulsion polymerisation.

The process of emulsion polymerisation consists of dispersing a water insoluble vinyl monomer phase with the aid of surfactants. Polymerisation is then carried out using the water phase as a heat sink to help the dissipation of the heat of reaction. The polymer formed is 'emulsified-in-situ' by adsorption into surfactant molecules as polymerisation proceeds and the resultant product is a homogeneous, relatively stable, emulsion of a high molecular weight polymer in water.

The process is extremely complex involving a minimum of four components and knowledge of the mechanism is still somewhat empirical.

ADVANTAGES OF THE EMULSION
POLYMERISATION TECHNIQUE

   i) High molecular weights with a narrow molecular weight distribution can be produced in a controlled reaction environment.

   ii) The emulsion polymer may be used in latex form (latex paints, etc.) or maybe precipitated from the emulsion and used in solutions (alcohol soluble printing inks, etc.), or neutralised to form a water dispersible resin (water based printing inks, etc.).

   iii) The latex viscosity is relatively low and is independent of the polymer molecular weight.

   iv) Water may be used as the sole solvent.

DISADVANTAGES

   i) When the polymer is not required in latex form it has to be precipitated from the emulsion and washed free from salts and surfactants.

   ii) Films formed from emulsion polymers have a tendency to be sensitive towards strong electrolytes.

iii) Emulsion latices have poor mechanical stability. Special methods are required for transferring and handling the products in bulk (e.g. low shear pumps, etc.).

PLANT
Similar plant to that used for suspension polymerisation would typically be used. Efficient stirring is essential otherwise destabilisation can occur. Economy of scale is important and large reactors with bulk storage are normally used. The use of gaseous monomers requires pressure reactors which is a specialised area. Thick reactor walls to withstand the high pressures involved when using gaseous monomers can lead to poor heat transfer characteristics for the reactor and formulations must take account of this. Surfactants are present and foaming particularly if there is a runaway exotherm, is an ever present problem. Correct reactor design can help but not alleviate the problem.

It is necessary for the reactor to have easy access for charging solid material if an in situ water phase is to be prepared. Unlike solution polymerisation the initiator(s), being in aqueous solution, is not mixed with monomer. Thus, separate addition tanks for monomer and initiator are required (possibly two separate header tanks). The rates of addition must be carefully controlled. The use of a pre-emulsified monomer mix requires an agitated header tank of fairly large capacity with possible steam heating. Pre-emulsions can be prepared in situ, or in a reactor, and then transferred to the agitated header tank.

PROCESS
The processes typically used for emulsion polymerisation are considered in detail later.

In essence monomer and initiator are added to a water phase consisting of surfactants, possibly colloids and additives. Temperatures vary between ambient for Redox initiation to $70-80°C$ for thermal initiation. Normally aliquots of monomer and initiator are added initially. Continuous addition then follows. Monomers can be emulsified in surfactant solutions to form a pre-emulsion. Whatever initiation technique is used the typical processing temperatures are $60-85°C$. Upon completion as determined by non-volatile content and the use of booster additions of either thermal or Redox initiators, the latex is cooled and filtered.

Filtration is different from that usually used for solution resins. A vibrating screen sieve is used and again care must be exercised to avoid foaming.

Releasing pressure can also cause foaming. Surfactants can stabilise foam which can be undesirable if an unexpected incident has occurred.

MECHANISM OF EMULSION POLYMERISATION
The mechanism of emulsion polymerisation still remains a matter for debate. There is no doubt that for model systems studied under ideal academic conditions with unrealistically low degrees of conversion (yields), theories may correlate with experiment. However, for industrial systems the commonly accepted mechanisms based upon the academic theories are somewhat empirical. One of the major differences is in the concentration of surfactants used and their method of addition. This has important consequences on micelle and critical micelle concentrations (C.M.C.) and the hypotheses

which depend upon these. Classical approaches to emulsion polymerisation mechanisms have been based on theories by Harkins and Smith & Ewart.

However, varying the water solubility of the monomers can cause different mechanisms to occur and if a pre-emulsion (surfactant mixed with monomers rather than in water phase) is used then all the theories are suspect.

Mechanisms of emulsion polymerisation are more complicated and the approaches are more varied than those for solution polymerisation.

All theories relate to the role played by surfactant forming micelles. Surfactants, which are amphoteric molecules are considered in detail in the next section. The surfactant molecule contains hydrophillic and hydrophobic groups, and these groups align themselves depending upon their surroundings. Surfactant molecules dissolve in water and also form micelles by the surfactant molecules orientating themselves, so that the hydrophillic groups are 'dissolved' in the water with the hydrophillic groups forced together in an agglomerate. The micelle can be depicted as follows and can be considered as an hydrated aggregate. Micelles are different to dissolved surfactant molecules.

Dissolved surfactant molecule

H_2O

H_2O H_2O

H_2O H_2O

Hydrophobic   Hydrophillic

more realistically H_2O H_2O

H_2O H_2O

A schematic representation of a micelle is:

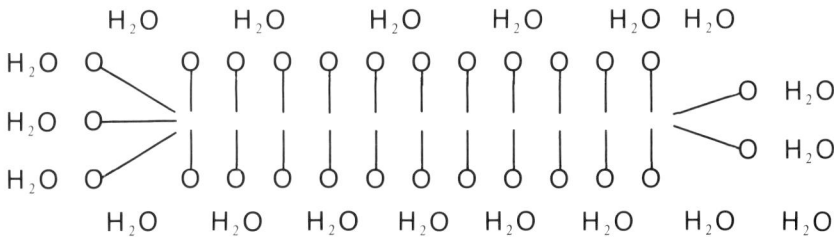

In droplets of monomer the hydrophobic (oil soluble) portion of the surfactant molecule 'dissolves' and the hydrophobic portion remains on the surface of the droplets thereby stabilising them. This can be schematically represented as follows:

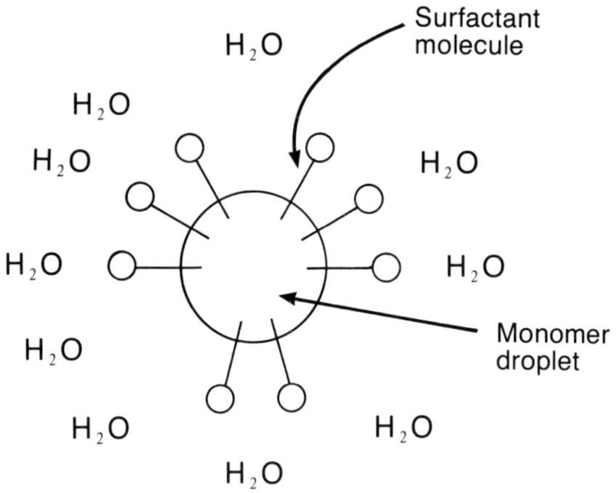

A simplified mechanism of emulsion polymerisation where monomer is added to a water phase containing only surfactant and initiator can be summarised as follows:

The water phase must be well agitated. Inadequate agitation will cause destabilisation of a latex, however good the theoretical stability.

The agitated water phase contains dissolved surfactant and micelles, the number and concentration of which depends upon the chemical structure and concentration of the surfactant, and the presence of other surfactants. Most industrially used latices contain more than one surfactant. It is known that micelles are formed because the surfactant has an excess solubility (apparent c.f. true solubility) and this excess is present in the form of micelles. Micelles generally consist of between 50–100 surfactant molecules.

There is a minimum concentration of emulsifier at which micelles first start to appear and this is referred to as the critical micelle concentration (CMC). The importance of the CMC is during polymerisation when the surfactant concentration is being depleted when the CMC determines the point at which micelles disappear.

The initial state is assumed to be that of a water phase consisting of micelles and dissolved surfactant.

Monomer is added to this system and the agitation breaks the monomer into droplets, the size partly depending upon the agitation and surface tension.

Besides being present in droplets, monomer dissolves in the water and is also solubilised in the surfactant micelles. Thus, monomer is present in three distinct states, even in the simplest of systems. With mixed emulsifiers, colloids and monomers of varying water solubilities, it is obvious that the system is even more complicated. The monomer droplets are stabilised by surfactant molecules as already depicted. The micelles contain hydrophobic centres and monomer is readily solubilised in them. These can be represented as follows, where M represents a monomer molecule.

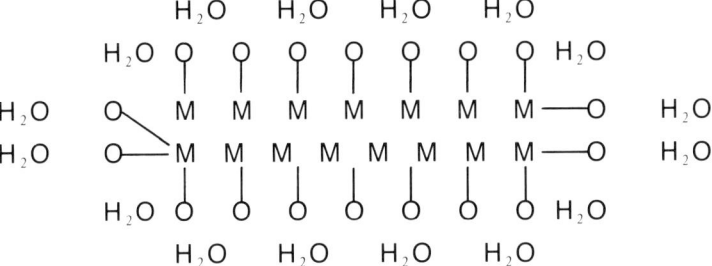

The affinity of a monomer for a particular surfactant is important. Different monomers require different surfactants.

The micelles containing monomer are significantly smaller than monomer droplets stabilised by surfactant and there are many more micelles.

The bulk of the initiator molecules are dissolved in the water. However, oil soluble initiators are sometimes used and these tend to dissolve in the monomer as droplets or micelles. This can cause further confusion. Thus, before any polymerisation occurs there is a dynamic equilibrium between the different parts of the system. The initiator dissociates into free radicals and these react with dissolved monomer forming a propagating species.

From this point theories differ as to the mechanisms. Smith & Ewart use three different classifications, depending upon the characteristics of the particular system. However, for the purposes of preparing industrial latices the following is probably as good a model as any to use assuming the required stability, conversion and film performance are obtained.

The growing species is then able to undergo a number of different mechanisms.

i) The growing species can absorb emulsifier and monomer becoming stabilised against flocculation, and able to grow until exhaustion of absorbed monomer which swells the polymer particle.

ii) The growing chain can be absorbed into a micelle containing monomer and a polymer particle (or nucleus) is formed which is stabilised by the surfactant of the micelle.

iii) Micelles containing monomer can absorb species outlined in i) before they have absorbed sufficient emulsifier to be stable.

iv) The growing species or free radical can enter a growing polymer particle (any outlined in i) to iii)). This results in chain termination.

v) The growing species can enter a particle containing polymer swollen with monomer but not propagating. The growing species initiates further polymerisation in the particle which is only terminated by absorption of further growing species.

vi) The depletion of dissolved and absorbed monomer causes monomer to diffuse from the monomer droplets to replenish the equilibrium.

vii) The surfactant is depleted by absorption into the polymer particles. A concentration is reached (CMC) where all the micelles disappear and further polymerisation is in the already formed particles. This point is typically claimed to be between 15−25% conversion of monomer to polymer.

At this point, some comments may clarify the situation.

The growing species enter micelles rather than monomer droplets because of the large number of micelles (c.f. droplets). Polymerisation of droplets leads to the formation of relatively low molecular weight polymer and also may be responsible for the formation of 'bits' or nibs (i.e. hard lumps of unstabilised polymer). The kinetics of addition polymerisation are obeyed. The process can be summarised by the following.

Initiation takes place in aqueous solution and is followed by further propagation in the aqueous phase. At some stage the growing oligomer (a polymer with only a few monomer units) radical becomes insoluble in water and requires stabilisation. This can be achieved by the radical entering a polymer particle, a monomer droplet or a surfactant micelle. Stabilisation can also occur by the radical adsorping onto a surfactant molecule in solution. It is this method which is the most important in the early stages of a polymerisation.

The stabilised oligomer radical can be regarded as a small polymer particle and it continues to grow as monomer molecules diffuse from aqueous solution into the particle. In the early stages of its growth the polymer particle requires continual adsorption of surfactant molecules to maintain its stability. Until it has a shell of surfactant molecules covering its surface, the particle will be unstable and can adsorb onto other polymer particles. As more polymer particles are formed, the rate of adsorption of new particles will increase until it equals their rate of formation. The period until this occurs is termed the seed stage.

Much of the controversy surrounding the mechanism is in the area of formation of particles. It is claimed that particle formation does not require micelles (Roe Ind. & Eng. Chem. *60*, (9), Sept. 1968, p. 20). Others claim that once the micelles are exhausted particle formation ceases.

The details of the theories are beyond the scope of this book and the reader is advised to consult specialist text books and published research papers to obtain further information.

It has been shown (by Smith & Ewart, *et al*) that for some systems the overall rate of emulsion polymerisation is proportional to the number of particles, the concentration of monomer and the propagation rate constant. But the number of particles is proportional to the concentration of emulsifier and initiator. The relationship to the initiator concentration is surprising.

The particle size is universally proportional to the number of particles and decreases exponentially with increasing emulsifier concentration (with constant surfactant type and initiator concentration) until a minimum value is reached. To reduce particle size further, other techniques may have to be used.

After the seed stage is completed the number of polymer particles remains constant and monomer diffuses into the particles; the number of free radicals in a particle can vary with values of 0 and 1 radicals per particle being most common. Because of the high polymer content of particles, commonly over 90%, the mass is a gel and the Tromsdorf effect controls termination within the particles. In other methods of polymerisation, the rate of termination is proportional to the square of the free radical concentration. However, in emulsion polymerisation, ignoring the Tromsdorf effect, the rate of termination is controlled by the rate at which radicals enter particles. This is proportional to the number of radicals. The rate of termination is both much lower than in other methods and also increases more slowly with increasing radical concentration. This leads to both higher molecular weights and faster reaction rates. During the main polymerisation period, monomer in monomer droplets diffuses into polymer particles, while surfactant in micelles and monomer droplets adsorps onto polymer-water interfaces. In synthetic rubber latices where a minimum of surfactant is used, all the surfactant is concentrated on the polymer surface. But in surface coating applications more surfactant is used and some micelles persist throughout the polymerisation.

To continue the discussion of emulsion polymerisation, let it be assumed that particles have been formed. The particles are swollen by monomer and as they grow, molecules of dissolved surfactant are absorbed in their surface micelles, which do not participate in particle formation breakdown to dissolved surfactant to maintain the equilibrium. These molecules are then available for future adsorption.

The number of particles is substantially less than the number of micelles originally present (e.g. 1 to 1000). The monomer droplets are gradually depleted as the particles grow. Initiation and termination can only occur when free radicals are absorbed into the particle. A commonly used term for the formation of the particle is nucleation, or particle nucleation because at about 20% conversion the particles are much smaller than their final size. Not only does surfactant stabilise droplets by a double (layer) charge effect and steric stabilisation, but the initiator often forms acidic fragments which solubilise and charge stabilise the growing species.

Monomer is absorbed, imbibed, by the growing particles until all the droplets have disappeared.

However, polymerisation still proceeds because the particles are swollen by monomer. It is only when traces of monomer remain that the rate of polymerisation decreases.

Some consequences of these mechanisms are:

a) Because initiation and termination depend upon radicals entering the particle, large molecular weights are obtainable. Grafting can also occur. This means that a polymer particle contains one or more free radicals. When a free radical enters the particle it either initiates chain propagation or terminates the propagating chain. This leads to long life time (relatively compared to other vinyl free radical polymerisation) radicals, and consequently large molecular weights and a stop/start polymerisation.

b) Particle size depends upon surfactant type and concentration (including pre-emulsion).

c) The use of a more water-soluble monomer will increase the rate of particle initiation (formation).

d) To remove residual monomer, an oil soluble initiator which will be absorbed into the particle is preferable. Alternatively, about 1% of a reactive water soluble monomer should be added with further initiator.

e) The particle size and surfactant system effect viscosity, particularly if colloid is present. The molecular weight of the polymer has no effect.

f) The surfactant molecules are an integral part of the polymer film and are often 'weak spots' for resistance properties, particularly aqueous solutions.

This weakening can be decreased by either cross-linking the latex which gives the best results, or by using as much copolymerisable surfactant as possible to replace other emulsifiers. Sodium vinyl sulphonate is an example.

The early theory of Smith & Ewart has dominated academic ideas but it has less influence on industrial thinking. The Smith-Ewart theory was developed to describe synthetic rubber emulsion where water insoluble monomers and a minimum of surfactant were used. Stage 1 of the polymerisation describes the formation of polymer particles by radicals entering micelles; when no further micelles exist, no new polymer particles can be formed. The number of particles is controlled by the amount of surfactant required to completely cover the surface of the particles and is proportional to the 0.6 power of the surfactant concentration. Stage 2 of the polymerisation involves particle growth and it is assumed that two radicals cannot co-exist within a polymer particle. The average number of particle radicals in a particle is 0.5 and the rate of polymerisation is proportional to the number of particles.

The weakness of the Smith-Ewart theory stems from its incorrect assumptions, especially in Stage 1. Firstly, emulsion polymers can be produced without surfactant by relying on the terminal sulphate groups to stabilise particles; this shows that micelles are not essential. Secondly, in many seed stages the number of particles reaches a constant value while surfactant micelles still exist. The third objection is that the theory takes no account of the marked differences found in both reaction rate and number of polymer particles obtained using different surfactants. In Stage 2 the Smith-Ewart theory has some validity, although its failure to take account of the Tromsdorf effect leads to minor errors. A more serious anomaly lies in the finding that, in some vinyl acetate polymerisations, the rate of polymerisation is proportional to the surface area of the particles; the reaction rate is dependent on the rate of diffusion of monomer into particles and not on chemical factors.

COMPONENTS USED IN
EMULSION POLYMERISATION

The four basic components of an emulsion polymer system are:
  i) Water.
  ii) Vinyl monomer – usually a mixture of comonomers.
  iii) Initiator – frequently more than one is used.
  iv) Surfactant – normally a mixture is used which frequently consists of non-ionic and ionic surfactants.

The other components normally present include:

a) Colloids.

b) Buffers.

Often post additions of the following are made when the coating is prepared to improve film formation and film performance characteristics.

c) Plasticisers.

d) Coalescing solvents.

The individual components are discussed separately in the section on components used in emulsion polymerisation.

A commonly used term by emulsion polymerisation technologists is 'water phase' which refers to the aqueous dispersion medium and includes water, surfactants, colloids and buffers. Depending upon the technologist, water phase may, or may not, be considered to include initiator or initiator systems.

Monomers

Most vinyl monomers will undergo emulsion polymerisation providing the monomer is easily available for reaction. This usually means that the monomer must be solubilised in the form of surfactant stabilised droplets or in micelles. Where the vinyl monomer contains long chain alkyl groups (e.g. lauryl and stearyl acrylates and methacrylates) polymerisation may be retarded resulting in low conversion of the monomer.

The rate of slow reactions can sometimes be increased by the addition of a small amount of water soluble solvent (e.g. glycols, alcohol and ketones) to the aqueous phase to improve the emulsification and hence the availability of the monomer for reaction.

Emulsion latices, film form by evaporation of the water phase followed by coalescence of the remaining polymer. For the polymer to form a coherent film it must be of sufficiently high molecular weight and also the temperature must be above M.F.F.T. (minimum film forming temperature) of the system. The M.F.F.T. (or M.F.T.) is governed by the Tg of the polymer and is normally somewhat lower than the Tg, due to the plasticising effect of the surfactant and any additives present.

Below the M.F.F.T. the film is chalky and non-continuous. At temperatures very much higher than the Tg, the film will be soft with poor adhesion and abrasion resistence. Most emulsion resins intended for ambient temperature applications have Tg's in the range of $0-30°C$.

Although it is possible to include additives specifically to reduce the M.F.F.T. these may leach out of the film, etc., and it is considered a much more desirable practise to control the Tg of the polymer by selection of the monomer type and composition. For this reason, most emulsion polymers are copolymers. A list of Tg's included with the monomer list may be found earlier in this chapter. In addition to Tg, other factors dictate the choice of monomer, such as the performance characteristics conferred by a particular monomer. A copolymer is designed to give a balance of properties in the final film including adhesion, flexibility, abrasion resistance, water resistance and ease of coalescence (film forming).

Typical 'hard' monomers include vinyl acetate where water often plasticises to give an M.F.F.T. about $10°C$ below the Tg.

Styrene and methyl methacrylate are the other two commonly used hard monomers. Both require significant amounts of flexibilising monomers. Both have advantages and disadvantages, and as a general rule, it is easier to prepare vinyl acetate emulsions rather than styrene, styrene acrylic or acrylic ones.

Vinyl chloride can also be considered as a hard monomer but a pressure reaction vessel is required. Vinyl chloride has many good resistance properties, but V.C.M. (vinyl chloride monomer) is highly toxic, limiting its use to specifically designed plants. Vinylidene chloride (V.D.C.) is a hard monomer which is sometimes substituted for vinyl chloride.

Flexibilising monomers are in effect internal plasticisers and include acrylic esters. The longer and more branched the aliphatic ester chain the more effective the plasticising efficiency for a given concentration. Typical esters are ethyl hexyl acrylate, butyl acrylate and ethyl acrylate. These monomers have to be used with styrene or methyl methacrylate. They can also be used with vinyl acetate as can vinyl versatate esters (Veova, ex. Shell). Veova 10 gives good alkali resistance to copolymers with vinyl acetate. Due to unfavourable reactivity ratios, Veova would not be used with styrene.

Butadiene has been used with styrene to form a large number of 'workhorse' latices for the coatings industry, but they tend to be for the less demanding applications (e.g. paper coatings (sizes)).

Ethylene is an excellent internal plasticiser particularly with vinyl acetate and vinyl chloride.

All commercial emulsion polymerisation requires the monomers to be added over a time period to enable adequate control of the exotherm.

Acid monomers such as acrylic and methacrylic acid, may be included to improve freeze-thaw stability. Ethylene is often used where good adhesion and flexibility is required (many emulsion adhesives are copolymers containing ethylene which acts as an internal plasticiser).

Amino monomers are often included as wet adhesion promoters.

Surfactants

The surfactant (and colloid) controls many of the properties of emulsion polymers and some knowledge of their nature and characteristics is a necessary pre-requisite to an understanding of this complex subject.

Any molecule containing a water soluble (hydrophilic or lyophobic) group chemically bound to a large water insoluble (hydrophobic or lyophilic) group will act as a surfactant. When a surfactant is dissolved in water, individual molecules concentrate at the water air interface where the hydrophilic groups are lying in the water and the hydrophobic group are sticking into the air. The surface tension of the solution is essentially that of the hydrophobic group which is about 30−40 dynes per cm. The concentration of individual surfactant molecules in water is limited and at a relatively low value surfactant molecules cluster together to form micelles. There is a critical concentration level below which a surfactant will not form micelles. The minimum level required for micelle formation is known as the 'critical micelle concentration' (C.M.C.) and varies with surfactant type. The critical micelle concentration is an important characteristic of surfactants. Surfactants are classified according to the ionic type of the hydrophilic group, whether anionic, cationic or nonionic. Most nonionic surfactants are made by reacting ethylene oxide with compounds containing an active hydrogen atom such as alcohols, carboxylic acids, phenols, amines or amides; the reaction occurs by a stepwise process:

$$ROH + CH_2 \overset{O}{-} CH_2 \longrightarrow ROCH_2 - CH_2OH$$

It is convenient to represent a nonionic surfactant by the name of the alcohol (or other active hydrogen compound) and the number of molecules of ethylene oxide reacted with it, e.g. lauryl alcohol +4 EO or nonyl phenol +15 EO. It should be remembered that the ethylene oxide content is an average figure.

The simplest anionic surfactants are the sodium or amine salts of fatty or rosin acids and these are used extensively for synthetic rubber emulsions. For the more stable emulsions required in surface coatings three main types are used: the first two are half esters of sulphuric or phosphoric acids, while the third utilises sulphonic acids where the sulphur atom is directly bonded to the lyophilic group and cannot be hydrolysed. All three species are normally used as their sodium salts. Cationic surfactants are not often used in emulsion polymers. The table below shows some surfactants commonly used in emulsion polymerisation and their critical micelle concentrations.

CRITICAL MICELLE CONCENTRATION OF SOME SURFACTANTS

Surfactant	C.M.C. (g per litre)
Lauryl alcohol + 6 EO	0.040
Nonyl phenol + 10 EO	0.026
Nonyl phenol + 15 EO	0.041
Nonyl phenol + 30 EO	0.12
Sodium lauryl sulphate	2.2
Lauryl alcohol + 2 EO sulphate, Na salt	1.1
Lauryl alcohol + 4 EO sulphate, Na salt	0.56
Sodium dodecylbenzene sulphonate	0.4
Sodium di-octyl sulphosuccinate	0.19

The role of a surfactant in emulsion polymerisation may vary during the course of the polymerisation.

Initially they contribute to the rate of polymerisation and particle formation. Once polymer is present the surfactant has to solubilise the polymer preventing precipitation, and when polymerisation is complete the surfactant is required to stabilise the emulsion preventing flocculation of the polymer and the formation of aggregates. Where the product is utilised in latex form the surfactant plays a major part in the performance characteristics and in particular:

i) Freeze thaw stability.

ii) Water sensitivity.

iii) Mechanical stability.

iv) Corrosion resistance.

v) Gloss.

The relationship between surfactant and particle size of the emulsion is of paramount importance. In general, increasing surfactant level decreases particle size. However, the relationship is an exponential one and each surfactant type has a maximum level above which particle size does not decrease.

Commercially available surfactants are rarely pure compounds. They often contain by-products of their formation reactions and there is often no direct replacement of one surfactant by a similar type of surfactant from a different manufacturer.

There is a wealth of literature on the various surfactants available. It is not intended to explore the relative merits of each type of surfactant in this chapter. Instead it is intended to give a brief background of the types available to enable the reader to have a basis for further reading. Comprehensive lists of surfactants detailing trade name and chemical type are available from British and American surfactant manufacturers, as well as from independent catalogues.

The development of a successful water phase for a polymer emulsion system can literally take years and much depends upon in-house experience. Successful water phase formulations are closely guarded secrets. It is more difficult to polymerise a bit free acrylic latex than a vinyl acetate one.

The selection of surfactants can be aided by considering the H.L.B. values for different surfactants. H.L.B. stands for Hydrophile Lypophile Balance. This enables a value to be given to the material to be emulsified and then to calculate a blend of emulsifiers with the same H.L.B. value. This method is not foolproof but it does enable initial guesses to be made. Experimentation is the only sure way to select the best water phase. H.L.B. values find more application in selecting emulsifiers for emulsifying resins which have already been formed, e.g. epoxy resin. Atlas Chemical Industries publish an excellent guide to selecting emulsifiers from an H.L.B. viewpoint.

It is important to realise that equivalents referring to surfactants for emulsion polymerisation have to be proven experimentally. Unlike many other products small changes in chemical structure can have a significant effect upon performance.

Surfactants may be divided into two types, ionic and non-ionic surfactants.

Ionic Surfactants

Ionic surfactants generally have a lower CMC than non-ionic surfactants. Ionic surfactants can be sub-divided into two types; a) cationic, and b) anionic types. Cationic types are not normally used in emulsion polymerisation. Anionic surfactants are often employed on their own, or in combination with non-ionic types, where their low CMC's provide low particle size emulsions. However, their use leads to a greater tendency for foaming and an inhanced water sensitivity in the final polymer film. It can also result in long term storage deficiencies. Chemically anionic surfactants are C_{12} to C_{18} carboxylic acid soaps, sulphonates, ethoxylated sulphonates, phosphates, sulphates and salts of polymeric acid derivatives.

Non-ionic Surfactants

The use of non-ionic surfactants may sometimes lead to the formation of small aggregates or grainy emulsions. This tendency is due to the weaker surface activity and the relative difficulty with which non-ionic surfactants form micelles.

As a result of these deficiencies non-ionic, anionic surfactant mixtures are normally employed in emulsion polymerisation. The ionic component allowing easy solubilising of the monomer whilst the non-ionic component confers emulsion polymer stability.

Chemically non-ionic surfactants are ethoxylated alkyl alcohols or alkyl phenols with an alkyl chain length of C_8 or greater.

The characteristics of the surfactant change with the degree of ethoxylation. Low degrees of ethoxylation tend to render the surfactant oil soluble whilst at higher levels the surfactant is water soluble. Most non-ionic surfactants used in emulsion polymerisation have a degree of ethoxylation corresponding to $15-30$ ethylene oxide units in the chain. At this level emulsion stability is enhanced compared to the lower ethoxylation levels.

Copolymerisable Surfactants

Copolymerisable surfactants are amphoteric molecules containing a reactive double bond. By participating in the polymerisation they improve the resistance properties of the film to water and aqueous solutions.

Pre-emulsions

The surfactant can not only be added to the water phase in the reactor. It is possible to emulsify the monomers in the surfactant and to continuously add this pre-emulsion to the water phase containing further surfactant.

This technique is frequently used for acrylic or styrene acrylic systems and is rare for vinyl acetate systems.

The pre-emulsion must be stable throughout the monomer addition period. The emulsifier system must be compatible with the emulsifiers in the reaction mixture. Depending upon whose theory is believed, it should be possible to generate new particles throughout the reaction. In practise it is only possible to produce commercially acceptable latices for some copolymer systems by using pre-emulsions. Often, only part of the anionic surfactant of the total emulsifier system is present in the pre-emulsion.

In production it is necessary to have agitated monomer addition tanks to enable pre-emulsions to be easily formed and stability maintained during their addition.

Selection of Surfactant

The range of surfactants is enormous and although many may be chemically similar it does not mean that they can be used as replacements without a full evaluation. Emulsion stability can be easily affected (particularly in borderline cases) by subtle differences in chemically similar surfactants. The selection of surfactants is aided by handbooks of surfactants with chemically similar types being listed.

Colloids

The term 'protective colloid' in an emulsion polymerisation refers to high molecular weight water soluble materials such as polyvinyl alcohol, cellulose derivatives and algenate.

Most water soluble macromolecules act as protective colloids in emulsion polymerisation. Macromolecule is a convenient term to denote a polymeric material whether prepared by polymerisation of a vinyl monomer or by polycondensation.

The colloids most commonly used in emulsion polymerisation are:

a) **Polyvinyl Alcohol**

This is produced by the partial hydrolysis of polyvinyl acetate.

b) **Cellulose Ethers**

Cellulose is a polyether which contains three hydroxyl groups per repeat unit. Methyl and hydroxyethyl ethers are widely used.

Other colloids sometimes used include poly(vinyl pyrrolidone), starch, styrene maleic anhydride copolymers and poly(acrylic acid).

The degree of hydrolysis of polyvinyl alcohols for emulsion polymerisation grades is about 85%. Completely hydrolysed material is insoluble in water due to intermolecular hydrogen bonding.

Polyvinyl alcohols formed from 85−90% hydrolysed polyvinyl acetate are used with vinyl acetate latices. The viscosity of the latex increases with increasing colloid content. Different grades give very different viscosities for the same weight. The higher the molecular weight of polyvinyl alcohol the higher the viscosity.

The grades of polyvinyl alcohol are referred to by number and/or letter. The higher the number, the higher the molecular weight. The letters used refer to different degrees of hydrolysis. K−80%; G−88%; A−98%; N−100%. The G grade is commonly used especially for polyvinyl acetate homopolymers. The water solubility depends upon the degree of hydrolysis, thus A and N grades are not readily soluble in water whilst K is. The less soluble the polyvinyl alcohol, the better the water resistance of the resulting polymeric film; surfactants and all other variables being similar.

The natural colloids of hydroxy cellulose and starch ethers find usage in all types of polymer latex and unlike polyvinyl alcohols there is much less variation in viscosity with molecular weights.

The hydroxy celluloses include methyl cellulose and especially hydroxy ethyl cellulose. The latter can be obtained in a variety of molecular weights and are most frequently used in vinyl acetate homopolymers and copolymers. Colloids are more effective than surfactants in protecting an emulsion polymer against aggregation. However, surfactants must also be present in order to form the micelles required during the polymerisation process. Where mixtures of colloid and surfactant are employed it is necessary to carefully balance the relative amounts of each species to avoid polymer flocculation. Emulsion latices are usually of low viscosity. Hydroxy ethyl cellulose (particularly the higher molecular weight grades) is often added as a thickener.

Colloids are frequently degraded by free radical attack during polymerisation. This improves compatibility and can lead to incorporation of colloid into the particle by graft copolymerisation. Colloids give a coarser particle size than colloid free systems.

Colloids are usually used with vinyl acetate copolymers, and for adhesives, high viscosity grades of polyvinyl alcohol are used. Colloids are not normally used in acrylics or styrene acrylic. If colloid is required for these systems then it would probably be a hydroxy ethyl cellulose, but pre-emulsification may be required.

Colloids play an important role in film formation and the properties of the finished film depend upon the cohesive strength and concentration of the colloid, and the degree of grafting.

Initiators

The initiators used in emulsion polymerisation are exclusively of the free radical type and of necessity they must be water soluble. The free radicals may be generated thermally or by use of a Redox couple.

Hydroperoxides are often used (e.g. t-butyl hydroperoxide) particularly as a post reaction initiator to scavenge the last vestages of unreacted monomer. This part of the process can be carried out in the maturing tank or blender, once the emulsion polymer has been discharged from the reactor, or the reactor can be utilised.

The major initiators used in commercial emulsion polymerisation are persulphates, with sodium, potassium and ammonium persulphate all being used. Whilst the initiating efficiency and half lives vary, ammonium persulphate is preferred because it is most soluble. The rate of dissociation of the initiator to give free radicals increases with temperature. When thermally generated free radicals are employed temperatures of $70-90°C$ are common. When Redox couples are employed the rate of free radical generation is increased (c.f. thermal initiation at that temperature) and temperatures as low as $30°C$ may be used.

Where persulphates are involved the major initiating radical is the sulphate ion radical.

$$^-O_3S-O-O-SO_3^- \longrightarrow 2\ S^\bullet O_4^-$$

Typical reducing agents used in the Redox couples are thiosulphates, metabisulphites and hydrosulphites. Frequently the Redox couple is combined with trace quantities of a ferrous salt providing an even more rapid generation of free radicals.

$$S_2O_8^{2-} + Fe^{2+} \longrightarrow S^\bullet O_4^- + SO_4^{2-} + Fe^{3+}$$

$$S_2O_8^{2-} + 2\ S_2O_3^{2-} \longrightarrow S_4O_6^{2-} + 2\ S^\bullet O_4^-$$

$$Fe^{3+} + 2\ S_2O_3^{2-} = \longrightarrow Fe^{2+} + \overset{\bullet}{S}_4O_6^{2-}$$

Buffers

Buffers are often added to stabilise pH because:

i) Some surfactants are pH sensitive with regard to micelle formation and latex stability.

ii) Some initiators are pH sensitive.

iii) Copolymerisation may occur better at a specific pH; e.g. acrylic acid and methacrylic acid do not form copolymers easily above pH 5.

iv) Some monomers maybe hydrolysed; e.g. polyvinyl acetate is hydrolysed at alkaline pH's.

Typical buffers are borax, sodium hydrogen phosphate and sodium bicarbonate salts.

Most latices are used at pH above 7.5. Where a buffer has been employed to keep the pH acidic during polymerisation it is often necessary to adjust the final pH upon completion of polymerisation. Ammonia is normally added to ensure an alkaline final product, but care must be exercised in its addition or destabilisation, or lumps may form. The pH of the latex depends upon the chemical nature of the polymer.

Persulphate initiators give acidic residues and buffers are frequently necessary to keep the pH of the system between four and five. Below pH 4 and above pH 9 vinyl acetate is susceptible to hydrolysis. Acrylic systems and particularly those containing acidic comonomers are unaffected by low pH's.

Water

The quality of the water used in emulsion polymerisation is important. The presence of polyvalent metal ions can cause instability. De-ionised water is preferred for the process and usually a limit of 250,000 ohm resistance is placed on the water used.

As well as acting as one phase of the emulsion the water also acts as a heat sink to remove the heat of polymerisation. Commercial processes are conducted with a water charge of between 35 and 65% of the total reactor charge and more frequently in the range of 40−50%.

Chain Transfer Agents

Chain transfer agents can be added to the monomer premix. The principles and chemical types are the same as those for solution polymerisation and they have been covered there. The main objective of a chain transfer agent is to control molecular weight by reducing the growing chain. This has been covered in the theory chapter.

External Plasticisers

Many of the commonly used external plasticisers can be used. As a general rule di-butyl phthalates are used with vinyl acetate and styrene is externally plasticised with phosphate esters or phthalates.

Coalescing Solvents

Glycols like ethylene and propylene are added to improve freeze thaw stability and rheological properties. In addition they also help in forming a continuous film, but not as much as lower evaporation rate glycols.

The solvents are considered to partition towards the hydrophilic network or polymer phase. Ethylene glycol is the former and ethylene glycol monobutyl ether acetate is the latter.

The influence and choice of co-adhesing solvents are considered in more detail in the film formation section.

THERMOPLASTIC EMULSIONS

These are commercially the most important class of vinyl and acrylic polymers for coatings and they are the basis of most decorative paints used today. The principal differences between solution and emulsion resins and processes were discussed earlier in this chapter, and the effect of the various monomers on the polymer properties was also detailed.

The most common type of emulsion for decorative paint is based on vinyl acetate internally plasticised with about 20 parts of vinyl versatate. Di-butyl or di-octyl maleate or fumarate are also used as plasticising monomers for vinyl acetate, but butyl acrylate or 2-ethylhexyl acrylate are now preferred. Vinyl acetate homopolymers externally plasticised with di-butyl phthalate are still widely used.

Styrene-acrylic copolymers have replaced vinyl versatate copolymers in some areas where improved properties were required such as hydrolysis resistance, but the highest performance coatings are now based on the more expensive full acrylic dispersions.

In recent years there has been a fast growth in the use of terpolymers based on vinyl acetate — ethylene — vinyl chloride. These terpolymers do not give quite as good performance properties as the acrylics, but ethylene and vinyl chloride are relatively inexpensive monomers.

THERMOSETTING EMULSIONS

Insignificant quantities of thermosetting emulsions are used compared with either thermosetting solution polymers or thermoplastic emulsions. This is because one of the big advantages of emulsion resins, i.e. very high molecular weight, is of much less importance for thermosetting systems where molecular weights are increased by cross-linking. However, emulsion resins having functional groups are commercially available. They are made by conventional emulsion resin techniques and their chemistry is as described in the section on thermosetting solutions.

The functional groups of some of the thermosetting emulsions can cause instability of the latex, either during processing or the final product. High levels of bit formation are not uncommon in thermoset latices.

FORMULATIONS AND METHODS FOR THE PREPARATION OF EMULSION POLYMERS

For convenience laboratory scale preparations are initially described.

Preparation of a Methyl Methacrylate — Homopolymer Latex

FORMULATION

De-ionised water	59.5
Sodium lauryl sulphate	0.8
Ammonium persulphate	0.1
Methyl methacrylate	39.6
	100.0

PROCEDURE

1. 550 g de-ionised water and 8 g sodium lauryl sulphate are added to a two litre flask fitted with a stirrer, reflux condenser, thermometer nitrogen inlet tube and two liquid addition inlets.
2. A solution of 1 g ammonium persulphate in 50 g de-ionised water is prepared and 20 g of this solution and 40 g methyl methacrylate monomer are added to the flask.
3. The stirrer and nitrogen flow are switched on and the flask slowly heated to 75°C.
4. When the contents of the flask assume a blue hue, the remainder of the initiator solution is added over a period of 90 minutes and 360 g methyl methacrylate are added over two hours. It is normal for such additions to be made using metering pumps to ensure reliable addition rates.
5. The emulsion is maintained at 75°C for 40 minutes after the end of the monomer addition to complete the reaction. The non-volatile content is determined and further initiator added if below theoretical, and the hold is continued for a further hour.
6. At the end of this period the nitrogen flow may be stopped, but stirring must be maintained during the cooling period to prevent surface drying of the emulsion. The product is a viscous white liquid with a bluish tinge; on viewing through the translucent emulsion, it appears red in colour.

Properties

Particle size 0.09 – 0.10 μ

Molecular weight c.a. 600,000

This example is the simplest formulation for emulsion polymerisation consisting of one material of each of the four essential components, i.e. water, one surfactant, one monomer and one initiator.

The apparatus used in this example is typical of a laboratory polymerisation, the flask would normally be heated in a water bath with thermostatic control. The blue hue is caused by the formation of very small particles which scatter blue light more than the longer wavelengths in the spectrum; red light which is scattered least, passes through the liquid giving a red appearance looking through the emulsion. The appearance of the blue colour which is often almost phosphorescent is a sign that polymerisation has started. Nitrogen must be used to remove oxygen from the system as the reaction is not carried out under reflux. The practice of adding only part of the initiator and of the monomer initially is commonly used.

Preparation of a Vinyl Acetate − MMA Copolymer Latex

FORMULATION

Vinyl acetate	40.2
Methyl methacrylate	13.4
Water	45.0
Sodium dodecylbenzene sulphonate	0.4
Ammonium persulphate	0.1
Ammonium acetate	0.1
Hydroxyethyl cellulose	0.8
	100.0

PROCEDURE

1. 100 g vinyl acetate, 25 g methyl methacrylate, 420 g water, 4 g sodium dodecylbenzene sulphonate, 1 g ammonium persulphate, 0.9 g ammonium acetate and 7.5 g hydroxyethyl cellulose are added to the flask and stirred.
2. The flask is heated until reflux starts with the temperature at about 67°C.
3. Refluxing is continued for 30 minutes when a mixture of 275 g vinyl acetate and 100 g methyl methacrylate is added over two hours.
4. Shortly after completion of the monomer addition the reflux temperature rose to 90−95°C indicating that reaction was complete.

Properties

pH of 4.8

Non-volatile content 55%

Viscosity 330 cps at 25°C

Particle size 0.15 μ

In this example a copolymer of VA/MMA is formed. A single surfactant, initiator, buffer and colloid are used again, keeping the components relatively straightforward.

Nitrogen is not required as vinyl acetate monomer maintains an oxygen free atmosphere above the reaction. The reactivity ratios for vinyl acetate and methyl methacrylate show that MMA is 20 to 40 times more reactive towards radicals than is vinyl acetate. The initial period of 30 minutes refluxing produces some copolymer rich in MMA and the addition of the balance of the monomer mixture at a rate approximating to that of polymerisation, ensures a random copolymer approximating to 73% by weight of vinyl acetate. The use of a colloid is partially responsible for the larger particle size compared to the previous example.

Preparation of a Vinyl Acetate Terpolymer Latex

FORMULATION

Vinyl acetate	9.400
Water	49.900
Sodium salt of octyl phenol + 4 EO sulphate	1.800
Ammonium persulphate	0.092
Tertiary butyl hydroperoxide	0.004
Sodium hydrosulphite	0.019
Sodium metabisulphite	0.085
Medium viscosity polyvinyl alcohol	0.900
Sodium salt of ethoxylated octylphenol sulphate	0.200
Ethyl acrylate	25.100
Methyl methacrylate	12.500
	100.000

PROCEDURE

1. 100 g vinyl acetate, 300 g water, 19 g of the sodium salt of octyl phenol + 4 EO sulphate, 0.18 g ammonium persulphate, 0.04 g tertiary butyl hydroperoxide and 0.2 g sodium hydrosulphite are added to a two litre polymerisation flask and stirred.

2. At the same time an initiator solution and a monomer emulsion are prepared. The initiator solutions contain 0.8 g ammonium persulphate and 0.9 g sodium meta-bisulphite each dissolved in 20 g water. 10 g of a medium viscosity polyvinyl alcohol and 2.5 g of the sodium salt of ethoxylated octylphenol sulphate are dissolved in 190 g water, then 267 g ethyl acrylate and 133 g methyl methacrylate are added with stirring to provide the monomer emulsion.

3. The contents of the flask are gently heated and when a sharp temperature rise to 78−80°C occurs, the monomer mixture and both initiator solutions are added over a period of two hours.

4. The temperature is maintained at 80°C throughout the addition and for a further hour before cooling.

Properties
pH 4−4.5
Non-volatile content 50%

This example is more complex and illustrates a more unusual method of preparing a terpolymer. However, because of the reactivity ratios of vinyl acetate with some monomers, alternative approaches are necessary. Colloid stabilisation and both Redox and thermal initiation are used.

The vinyl acetate is essentially polymerised before the acrylate monomers are added and a random copolymer cannot be produced. The conditions are selected to induce grafting of the acrylate onto the vinyl acetate backbone. The use of a colloid to emulsify the acrylic monomers discourages the formation of new particles and the acrylates therefore polymerise in the vinyl acetate particles. The use of higher initiator levels than in the original homopolymerisation produces more radicals and hence a higher probability of grafting.

Preparation of a Vinyl Acetate Homopolymer Latex stabilised with Polyvinyl Alcohol

FORMULATION

Polyvinyl alcohol (medium viscosity) (88% hydrolysis)	1.57
Water	44.22
Vinyl acetate	54.05
Sodium acetate	0.1
Ammonium persulphate	0.06
	100.00

PROCEDURE

1. 16 g medium viscosity grade of polyvinyl alcohol with 88% hydrolysis are stirred into 450 g water in a two litre polymerisation flask.

2. The solution is heated to 90°C with stirring and held at that temperature for 30 minutes before cooling to 30°C.

3. 50 g vinyl acetate, 1.0 g sodium acetate and 0.6 g ammonium persulphate are added and the whole heated slowly to 70°C. At 67−68°C reflux starts with severe foaming; it may be necessary to remove the water bath for a few minutes to prevent foam shooting out through the condenser. The start of polymerisation is signalled by the sudden collapse of the foam with the simultaneous formation of a blue tinge to the reaction mixture.

4. The temperature is raised to 75°C and the addition of 500 g vinyl acetate is started. This is timed to take 90 minutes.

5. At the end of the monomer addition the bath temperature is raised to 98°C and the reaction temperature increases to 92−93°C in 10−15 minutes.

6. After holding for ten minutes at this temperature the product is cooled. The product is a milky white viscous liquid with a pH of 4.5. The viscosity is 20,000 cps at 55% non-volatile content and the particle size is about 0.75 micron.

```
┌─────────────────────────────────────────────────┐
│  Properties                                     │
│  pH 4.5                                          │
│  Non-volatile content 55%                        │
│  Viscosity 20,000 cps at 25°C                    │
│  Particle size 0.75 μ                            │
│  Appearance milky white viscous liquid           │
└─────────────────────────────────────────────────┘
```

This formulation illustrates a colloid stabilised polyvinyl acetate latex. The colloid stabilises the latex obviating the need for surfactant. This is unusual. The persulphate breaks down into acidic groups which also assist in stabilising the latex. Non VA latices need surfactant for stability.

The process of 'cooking' the polyvinyl alcohol affects the emulsion properties. Cooking probably decreases intermolecular hydrogen bonding in the polyvinyl alcohol; increased cooking leads to higher emulsion viscosities and to greater grafting of vinyl acetate onto the polyvinyl alcohol. Foaming of emulsion polymers always occurs with polyvinyl alcohol at the start of a reaction. One technique to overcome this problem is to carry out the reaction under pressure when no reflux occurs. However, this solution introduces another problem, that of cooling a viscous emulsion without the use of the latent heat of vaporisation.

During the final heating period, the reaction mixture has a sharp rise in viscosity caused by particle coagulation. Commonly, the particle size of a polyvinyl alcohol grade increases slowly during the monomer addition period to about 0.30 micron.

The final heating period sees a sharp rise to about one micron and then a small decrease to the value of 0.75. The final decrease is probably due to a smoothing out of the coagulated particles. The particle agglomeration process is reproducible and characteristic of polyvinyl alcohol grades. The rapid reaction of 105 minutes compared with the 3−5 hours for solution polymerisation is a major advantage of the emulsion method.

Preparation of a Vinyl Acetate Homopolymer Latex stabilised with Hydroxyethyl Cellulose − Thermal Initiation

FORMULATION

Water	43.49
Nonyl phenol + 35 EO	2.07
Nonyl phenol + 9.5 EO	1.03
Sodium acetate	0.04
Hydroxyethyl cellulose	1.55
Vinyl acetate	51.76
Ammonium persulphate	0.06
	100.00

PROCEDURE

1. 400 g water, 20 g nonyl phenol + 35 EO, 10 g nonyl phenol + 9.5 EO and 0.4 g sodium acetate are added to a two litre flask and stirred.
2. 15 g hydroxyethyl cellulose are added slowly and the mixture stirred until all the colloid has dissolved.
3. 75 g vinyl acetate and 0.4 g ammonium persulphate are added and the whole heated to reflux maintaining the water bath at 75°C.
4. When the flask contents reach 75°C the addition of 425 g vinyl acetate and a solution of 0.2 g ammonium persulphate in 20 g water is started, timed over 140 minutes.
5. When the monomer addition is completed, the temperature is raised to 80°C and held for 30 minutes before cooling.

Properties

pH 4.7

Non-volatile content 54%

Viscosity 1500 cps at 25°C

Particle size 0.45 μ

This and the next example illustrate the difference between thermal and Redox initiation. Mixed surfactants, colloid and buffer are present. The formulation is restricted to homopolymer to illustrate the differences in the other parts of the formulation.

Compared with polyvinyl alcohol grades, the use of hydroxy ethyl cellulose provides no processing difficulties and gives products of much lower viscosity. The longer reaction time is due in part to reaction between the colloid and sulphate ion radicals. This occurs at the cellulosic ether bridge leading to a hydroxyethyl cellulose sulphate and a cellulosic free radical which then initiates polymerisation of the vinyl acetate. By keeping a low concentration of initiator, this reaction is minimised and the viscosity drop it causes is also reduced. The conversion is only 95% which is a low value for an emulsion polymer.

Preparation of a Vinyl Acetate Homopolymer Latex stabilised with Hydroxy Ethyl Cellulose — Redox Initiation

FORMULATION

Water	43.38
Nonyl phenol + 35 EO	2.06
Nonyl phenol + 9.5 EO	1.03
Sodium acetate	0.04
Hydroxyethyl cellulose	1.55
Vinyl acetate	51.63
Hydrogen peroxide (30%)	9.21
Sodium thiosulphate in water	0.10
	100.00

PROCEDURE

1. The previous example is repeated replacing the ammonium persulphate with a Redox initiator.
2. 1 g of 30% hydrogen peroxide is added initially and two solutions are added slowly. These are 1 g hydrogen peroxide in 20 g water and 1 g sodium thiosulphate in 20 g water.
3. Reaction only starts when the thiosulphate solution is added and the monomer addition is delayed until the blue colour is visible and then added over two hours.

Properties

pH 4.9

Non-volatile content 54.9%

Viscosity 1300 cps at 25°C

Particle size 0.47 μ

The conversion is improved to over 99% but the other emulsion properties are not significantly changed. The detailed chemistry of Redox reactions is not always known but the principle may be illustrated by the reaction between ferrous iron and hydrogen peroxide:

$$Fe^{2+} + HOOH \longrightarrow Fe^{3+} + OH^- + \cdot OH$$

Commonly used reductants in Redox mixtures are bisulphites, thiosulphates, di-thionites, ferrous salts and ascorbic acid. Either hydrogen peroxide or persulphates may be used as the oxidising component. Redox mixtures can generate free radicals more rapidly than the peroxyl compound at the same temperature, but the efficiency of radical production is under half of that of the thermal decomposition of the oxidant. Redox systems are normally only used for low temperature polymerisations or for reactions with special difficulties.

Preparation of a VA−Veova Copolymer Latex

FORMULATION

Sodium dodecylbenzene sulphonate	0.26
Nonyl phenol + 15EO	0.87
Sodium bicarbonate	0.17
Water	42.98
Hydroxyethyl cellulose	1.22
Vinyl acetate	46.03
Veova 911	8.25
Ammonium persulphate	0.22
	100.00

PROCEDURE

1. 3g sodium dodecylbenzene sulphonate, 10g nonyl phenol + 15 EO, 2g sodium bicarbonate and 425g water are added to a two litre flask; the contents are stirred and 14g hydroxyethyl cellulose are added slowly.

2. 530g vinyl acetate and 95g Veova 911 are mixed.

3. 60g of this mixture and 1.5g ammonium persulphate are added to the reaction flask which is then heated to 75°C. When the initial reflux has ceased, the monomer mixture is added over a period of four hours.

4. 30 minutes after the start of the monomer addition, a solution of 1g ammonium persulphate in 70g of water is added timed for 3½ hours. At the same time the reaction temperature is increased to 80°C.

5. The emulsion is held at 80°C for 45 minutes after the end of the monomer addition before cooling.

> **Properties**
>
> pH 4.5
>
> Non-volatile content 56.3%
>
> Viscosity 2100 cps at 25°C
>
> Particle size 0.35 μ

Anionic and non-ionic surfactants are mixed. A buffer and colloid are also present. Thermal initiation is used. The VA—Veova copolymer is internally plasticised.

Veova 911 is a mixture of esters of vinyl alcohol with branched chain acids containing nine or eleven carbon atoms. It is a softening comonomer for vinyl acetate with which its reactivity ratios are almost unity so that random copolymers are readily produced. The use of some anionic surfactant reduces the particle size. It is the smaller particle size which is the principal factor in the viscosity increase as compared with the two previous examples.

Note the high conversion (99½%) produced by high initiator levels and long reaction times.

Preparation of a VA—2-Ethylhexyl Acrylate Copolymer Latex

FORMULATION

Polyvinyl alcohol (medium viscosity) (88% hydrolysis)	1.290
Water	44.060
Nonyl phenol + 12 EO	0.640
Ammonium acetate	0.053
Ammonium persulphate	0.107
Vinyl acetate	42.980
2-ethylhexyl acrylate	10.740
Sodium dodecylbenzene sulphonate	0.130
	100.000

PROCEDURE

1. 12 g of a medium viscosity polyvinyl alcohol of 88% hydrolysis, dissolved in 370 g water, are cooked at 80°C for one hour and cooled.
2. 6 g nonyl phenol + 12 EO, 0.5 g ammonium acetate, 0.5 g ammonium persulphate and 20 g vinyl acetate are added, and the whole heated to reflux.

3. As soon as the foam breaks, addition of the monomer is started, timed over four hours. The monomer is a mixture of 100 g 2-ethylhexyl acrylate with 380 g vinyl acetate. At the same time the addition of a solution of 0.5 g ammonium persulphate and 1.2 g sodium dodecylbenzene sulphonate in 40 g water is started, also timed for four hours.

4. The temperature is held at 75°C for a further 30 minutes after the end of the monomer addition when the emulsion is cooled.

Properties

pH 4.4

Non-volatile content 55.5%

Viscosity 9000 cps at 25°C

Particle size 0.65 μ

In this example, 2-ethylhexyl acrylate is the internal plasticising comonomer. Mixed surfactants (anionic and non-ionic), buffer and colloid are present.

Random copolymerisation between the two monomers is achieved by the slow addition of monomer mix to a vinyl acetate rich reaction; the slower reaction is partially due to the lower solubility of 2-ethylhexyl acrylate in water. The use of an anionic surfactant is required to assist the incorporation of 2-EHA and results in a slight reduction in particle size.

Preparation of a VA − Di-butyl Maleate Copolymer Latex

FORMULATION

Lauryl alcohol + 20 EO	2.35
Polyvinyl pyrrolidone (molecular weight 300,000)	0.56
Water	49.88
Potassium persulphate	0.14
Di-butyl maleate	14.12
Vinyl acetate	32.95
	100.00

PROCEDURE

1. 25 g lauryl alcohol + 20 EO and 6 g polyvinyl pyrrolidone (molecular weight 300,000) are dissolved in 500 g water and heated to 80°C.

2. 10g of a solution of 1.5g potassium persulphate in 30g water and 20g of a mixture of 150g di-butyl maleate with 350g vinyl acetate are added to the hot solution.

3. The balance of both mixtures are added at a constant rate over 4½ hours. The reaction is maintained at 80°C throughout the monomer addition and for a further 30 minutes before cooling.

Properties

pH 4.3

Non-volatile content 50.5%

Viscosity 250 cps at 25°C

Particle size 0.6 μ

Di-butyl maleate is copolymerised with vinyl acetate. Polyvinyl pyrrolidone (pvp) is the colloid. Thermal initiation is used.

Polyvinyl pyrrolidone gives low viscosity emulsions when used as the only colloid and the emulsion tends to have poor stability. The great advantage of pvp is its compatibility with many polymers which give clear integral films.

Preparation of a Styrene Shellac Latex

FORMULATION

50% solution bleached lac in sodium carbonate (pH 7.3 – 7.5)	30.14
Water	43.93
Ammonium persulphate	0.09
Styrene	25.84
	100.00

PROCEDURE

1. 350g of a 50% solution of bleached lac in sodium carbonate with a pH of 7.3–7.5, 450g water, 0.5g ammonium persulphate and 40g styrene are heated to 75°C with stirring.

2. After ten minutes at that temperature, addition of 260g styrene and a solution of 0.6g ammonium persulphate in 60g water are added over 2½ hours.

3. The reaction mixture is held at 75°C for 45 minutes after the end of the monomer addition before cooling.

```
┌─────────────────────────────────────────┐
│  Properties                             │
│  pH 7.3                                 │
│  Non-volatile content 40%               │
│  Viscosity 600 cps at 25°C              │
│  Particle size 0.11 μ                   │
└─────────────────────────────────────────┘
```

pH control is the key to this rather straightforward example. At a pH above 7.5 the rate of decomposition of the persulphate ion is too rapid, and many radicals react to form sulphuric acid and oxygen. At pH below 7.1, the stabilisation of the shellac solution is reduced and poor stability results.

Preparation of an Acrylate Copolymer Latex containing Acidic Groups

FORMULATION

De-ionised water	63.79
Sodium lauryl sulphate	1.03
Methyl methacrylate	16.38
2-ethylhexyl acrylate	16.38
Acrylic acid	1.72
Ammonium persulphate	0.35
Sodium bisulphite	0.35
	100.00

PROCEDURE
1. 740 g de-ionised water, 12 g sodium lauryl sulphate and 150 g of a monomer mixture (comprised by 190 g methyl acrylate, 190 g 2-ethylhexyl acrylate and 20 g acrylic acid), are added to a two litre flask and purged with nitrogen.
2. After 30 minutes purging, 4 g each of ammonium persulphate and sodium bisulphite are added and slow heating applied.
3. When the temperature reaches 30°C the remainder of the monomer mixture is added over a period of 105 minutes. During the monomer addition the temperature is gradually increased to 60°C, and then it is raised rapidly to 90°C and held for 30 minutes before cooling.

```
Properties
pH 3.9
Non-volatile content 35%
Viscosity 450 cps at 25°C
Particle size 0.13 μ
```

In this example the use of a Redox initiator to start the polymerisation at a low temperature is illustrated. Previous examples have described isothermal polymerisations, but this reaction pattern is commonly used in industry as it reduces the need for cooling. On a large reactor, the heat required to raise the temperature would be provided by polymerisation. The technique has been termed 'shotgun polymerisation'. The small particle size and an increase in particle hydration due to the acid groups limit the non-volatile content to 35%.

Preparation of an Acrylate-Vinylidine Chloride Copolymer Latex

FORMULATION

Water	55.26
Sodium lauryl sulphate	0.59
Nonyl phenol + 20 EO	1.10
Butyl acrylate	21.25
Vinylidine chloride	21.25
Hydrogen peroxide 30%	0.13
Ascorbic acid	0.42
	100.00

PROCEDURE

1. 550 g water, 7 g sodium lauryl sulphate and 13 g nonyl phenol + 20 EO are stirred in a two litre flask which is purged with nitrogen. In this preparation, the reflux condenser is fed with water at 1−5°C through a circulating pump.

2. After purging for 20 minutes, 30 g of monomer mixture are added and the rate of nitrogen flow reduced to a low level. The monomer mixture is made from 250 g each of butyl acrylate and vinylidine chloride. At the same time the top end of the condenser is attached to two wash bottles in series, each containing acidified potassium permanganate solution to trap any vinylidine chloride vapour.

3. The flask is heated to 36–38°C and two Redox solutions are slow added, timed over five hours. The Redox solutions are made by dissolving 1.5 g 30% hydrogen peroxide in 50 g water and 5 g ascorbic acid also in 50 g water.

4. The addition of the balance of the monomer mixture is started as soon as polymerisation starts and is timed for four hours. When the Redox addition is complete the product is cooled.

Properties

pH 3.3

Non-volatile content 44%

Viscosity 150 cps at 25°C

Particle size 0.17 μ

Vinylidine chloride is very volatile (boiling point 38°C) and toxic. The use of very cold water for the condenser is essential to prevent significant loss of monomer carried over by the nitrogen stream. The use of a closed system means that a low flow rate of nitrogen can be used during the reaction and the permanganate solution should react chemically with any vinylidine chloride that does escape from the system.

The Redox system of ascorbic acid and hydrogen peroxide is an extremely rapid source of free radicals even at low temperatures; in the preceding example the Redox pair of persulphate and bisulphite continues to produce radicals after 2½ hours, but for the peroxide – ascorbic system the reaction is complete in a few seconds. If one component were to be added initially and the other added slowly over a period, very little polymerisation would occur because all the added component would react almost instantaneously, and the free radicals would neutralise each other before they could initiate polymerisation.

It is important that both components should be added separately and preferably at widely spaced locations in the flask. Ascorbic acid always gives low pH and even after neutralisation it tends to cause poor colour stability in polymer films. Commercial production of this recipe would use a closed system and the temperature would be increased as far as the pressure in the system would allow to improve cooling efficiency.

Preparation of a Vinyl Chloride-Ethylene Copolymer Latex

FORMULATION

Di-potassium hydrogen phosphate	0.14
Potassium persulphate	0.60
Ammonium salt of a styrene-maleic anhydride copolymer in water (30%)	2.63
Vinyl chloride	25.24
Ethylene	7.57
Water	63.82
	100.00

PROCEDURE

1. 4 g di-potassium hydrogen phosphate, 17 g potassium persulphate, 25 g of a 30% aqueous solution of the ammonium salt of a styrene-maleic anhydride copolymer and 1820 g water were added to a five litre stainless steel autoclave.
2. The autoclave is flushed by adding ethylene at 20 atmospheres pressure, three times, and 240 g vinyl chloride were pumped in. Ethylene is added to give a pressure of 200 atmospheres, and the autoclave heated to and maintained at 70°C.
3. After 3½ hours, 240 g vinyl chloride and 25 g 30% SMA solution were pumped in over 30 minutes.
4. A further addition of the same quantities of both materials is made after 90 minutes.
5. The ethylene pressure is kept at 180−220 atmospheres throughout.
6. Sixty minutes after the last additions, the pressure is (gradually) reduced and the whole cooled.

Properties

pH 6.1

Non-volatile content 33%

Ethylene content of polymer 30%

In high pressure polymerisation the ethylene is not weighed nor metered into the reactor. A relatively constant pressure of the olefin is maintained, the magnitude of the pressure and the time of reaction determining the amount of ethylene reacted. The very low solubility of ethylene in water gives slow diffusion into polymer particles and relatively long reaction times are common.

Preparation of a Vinyl Acetate-Ethylene Copolymer Latex

FORMULATION

Water	45.054
Nonyl phenol + 35 EO	1.410
Nonyl phenol + 10 EO	0.845
Sodium salt of vinyl sulphonic acid	0.282
Sodium lauryl sulphate	0.084
Citric acid	0.113
Di-sodium hydrogen phosphate	0.056
Vinyl acetate	39.430
Potassium persulphate	0.676
Ethylene	8.670
Sodium formaldehyde sulphoxylate (4% solution)	3.380
	100.000

PROCEDURE

1. 800 g water, 25 g nonyl phenol + 35 EO, 15 g nonyl phenol + 10 EO, 5 g sodium salt of vinyl sulphonic acid, 1.5 g sodium lauryl sulphate, 2 g citric acid, 1 g di-sodium hydrogen phosphate and 700 g vinyl acetate are added to a five litre autoclave.
2. After purging with nitrogen, 12 g potassium persulphate were added and the autoclave heated to 50°C while ethylene was pumped in to give a pressure of 35 atmospheres.
3. 60 g of a 4% solution of sodium formaldehyde sulphoxylate were added over a period of 4½ hours, during which the ethylene pressure was maintained at 35 atmospheres.
4. The pressure was released and the product cooled.

Properties
pH 4.2
Non-volatile content 48%
Particle size 0.21 μ
Ethylene content of polymer 18%

The vinyl sulphonic acid salt will provide some stabilisation by copolymerising with the vinyl acetate, but the majority of the salt will homopolymerise to form a colloid. Sodium formaldehyde sulphoxylate is made by reacting sodium bisulphite with formaldehyde and is often used in Redox mixtures. It is generally more reactive than sulphites, but not so rapid as ascorbic acids.

Mixed non-ionic and ionic surfactants are used. Di-sodium hydrogen phosphate acts as a buffer.

Preparation of an Acrylamide-Vinyl Acetate-Ethylene Terpolymer Latex

FORMULATION

Vinyl acetate	46.505
Nonyl phenol + 15 EO	1.094
Sodium dodecyl sulphonate	0.410
Acrylamide	0.301
Hydroxyethyl cellulose (0.35% aqueous solution)	0.164
Ammonium persulphate	0.096
Water	49.240
Ethylene	2.190
	100.000

PROCEDURE

1. An emulsion is prepared by adding 1700 g vinyl acetate to a solution of 40 g nonyl phenol + 15 EO, 15 g sodium dodecyl sulphonate, 11 g acrylamide and 6 g hydroxyethyl cellulose in 1750 g water.

2. The initiator solution is prepared by dissolving 3.5 g ammonium persulphate in 50 g water.

3. 1000 g of the monomer emulsion and 10 g of the initiator solution are added to a five litre autoclave, which was purged of oxygen by repeated purging with nitrogen. Ethylene forms explosive mixtures with air. The nitrogen is then removed by flushing the reactor with ethylene.

4. The autoclave is filled with ethylene to a pressure of ten atmospheres and heated to 78°C.

5. After 15 minutes, addition of initiator and monomer emulsion is started at a rate to give a five hour addition. At the end of each hour the addition rate for both components is reduced by 10% giving a seven hour addition.

6. The reaction mixture was allowed to react for a further hour with the ethylene pressure being maintained at ten atmospheres throughout. The copolymer had an ethylene content of 4.5%.

The preparation was repeated using different ethylene pressures, and the following results were obtained.

Ethylene pressure	% of ethylene in copolymer
10 atmospheres	4.5
15 atmospheres	6.8
20 atmospheres	8.9
25 atmospheres	12.6
30 atmospheres	14.8
50 atmospheres	26.9
150 atmospheres	45.7
300 atmospheres	63.2

Mixed surfactants and colloid are used. Acrylamide copolymerises and is a possible site for cross-linking. Frequently it is necessary to gently warm the monomer mixture to dissolve acrylamide. A similar formulation without acrylamide could be used for an ethylene, vinyl acetate copolymer. Some production scale examples of emulsion polymerisation follow.

Preparation of an Acrylic Copolymer Emulsion for Textile Sizing

FORMULATION

Aqueous Phase	
Water	24.00
Sodium metabisulphite	0.13
Octyl phenol ethylene oxide (30 EO)	2.00
Monomer Phase	
Water	18.54
Octyl phenol ethylene oxide (30 EO)	0.86
Sodium lauryl sulphate	1.87
Sodium persulphate	0.13
Ethyl acrylate	50.87
n-methylol acrylamide (48% solution)	1.08
Methacrylic acid	0.52
	100.00

PROCEDURE

1. Charge aqueous phase to reactor. Start stirrer and heat to 65°C.
2. At 65°C add 10% of the premixed and pre-emulsified monomer phase.
3. Allow the reaction exotherm to take the temperature to 70−75°C.
4. Hold at 70−75°C for 30 minutes then begin addition of the remainder of the premixed monomer phase over four hours. Maintain the temperature at 70−75°C throughout the addition period.
5. Thirty minutes after the addition is complete, add a 'spike' of sodium metabisulphite, followed by an equal quantity of sodium persulphate.
6. Hold for one hour at 75°C and then discharge to a maturing tank.
7. Stir overnight in the maturing tank at ca 35°C, adding further 'spikes' of Redox initiator if required, to obtain free monomer contents (determined by G.L.C. measurements) below 0.5%.
8. Strain via a vibrating filter.

Preparation of a Vinyl Acetate
Copolymer Latex (Internally Plasticised)

FORMULATION

Aqueous Phase	
Water	44.90
Lauryl alcohol ethoxylate	0.60
Borax	0.20
Hydroxy ethyl cellulose	2.70
Nonyl phenol ethylene oxide	1.40
Sodium persulphate	0.20
Monomer Phase	
Vinyl acetate	46.00
Vinyl versatate	4.00
	100.00

PROCEDURE

1. Charge reactor with aqueous phase. Stir and adjust pH to 4.5−5.5 with bicarbonate solution if required.
2. Heat to 65°C.
3. At 65°C add 10% of premixed monomer phase and allow the temperature to rise to 75°C.

4. Hold at 75°C for 30 minutes and then add remainder of monomer phase over three hours at 75°C.
5. When addition is complete allow the temperature to increase to 80°C.
6. Hold at this temperature for two hours and then discharge to a maturing tank.
7. Add a spike of equal amounts of sodium persulphate and sodium metabilsulphate, and stir overnight at 35°C.
8. Strain via a vibrating filter.

Preparation of a Styrene-Acrylic Emulsion (with pre-emulsion)

FORMULATION

Reactor Charge	
Water	18.35
Stabilising colloid	0.25
Ammonium persulphate	0.20
Ammonium hydroxide (25%)	0.50
De-foamer	0.05
Bacteriocide	0.10
Fungicide	0.05
Pre-emulsion Tank	
Water	30.00
Styrene	23.00
Butyl acrylate	24.00
Methacrylic acid	1.00
Surfactant	0.20
Colloid	2.00
Di-methylethanolamine	0.30
	100.00

PROCEDURE

1. Charge the monomers to the pre-emulsion tank and stir in all the additives. Stop agitation and add the water.
2. Mix for five minutes and then allow to settle for 15 minutes.

3. Repeat till pre-emulsion is homogeneous.
4. Charge water to reactor and heat to 80°C while purging with inert gas.
5. Add the persulphate, ammonium hydroxide and de-foamer, and immediately begin to run in the pre-emulsion mix.
6. Complete the addition of the pre-mix over a two hour period while holding the temperature between 80−85°C.
7. Maintain at 85°C for a further two hours.
8. Cool to 50°C, and add fungicide and bacteriocide.
9. Discharge through 60 micron vibrating screen.

Preparation of a Thermosetting Acrylic Emulsion

FORMULATION

Reactor Charge	
Water	15.0
Stabilising colloid	1.0
Ammonium persulphate	0.1
Pre-emulsion Tank Charge	
Water	44.9
Styrene	21.0
n-butylacrylate	12.0
Hydroxyethyl methacrylate	5.0
Emulsifier	0.5
Colloid	0.5
	100.0

PROCEDURE
1. Mix the monomers and surface agents in the pre-emulsion tank, and then stir in the water and make the monomer emulsion.
2. Charge the reactor and run in 10% of the monomer premix.
3. Heat to 80°C while stirring, and run in the rest of the monomer premix over a 2½ hour period.
4. Stir and heat at 90°C for two hours.
5. Discharge at 50°C through 60 micron shaking screen.

Preparation of a Water Soluble Acrylic Resin via Emulsion Polymerisation

FORMULATION

Reactor Charge	
Water	18.00
Stabilising colloid	0.25
Butyl mercaptan	0.50
Isopropanol	1.50
Ammonium persulphate	0.50
Ammonium hydroxide (25%)	0.50
De-foamer	Trace
Pre-emulsion Tank Charge	
Water	30.00
Acrylic acid	9.00
Ethyl acrylate	26.00
Butyl methacrylate	12.00
Surfactant	0.25
Stabilising colloid	1.50
	100.00

PREPARATION OF EMULSION POLYMER

1. Charge pre-emulsion tank and prepare pre-emulsion at ambient temperature.
2. Charge reactor, stir and heat to 80°C.
3. Run pre-emulsion gradually into the reactor over a two hour period.
4. Hold under reflux for further two hours.
5. Check that solids content is correct and allow reactor to cool.

It is now possible to neutralise the pendant acid groups, thereby solubilising the polymer. Care must be exercised during the addition of the basic compound to avoid bit formation, etc.

PREPARATION OF POLYMER SOLUTION

1. Stir 30 parts of water into the 100 parts of acrylic emulsion in the reactor.
2. Slowly add tri-methylamine while stirring till pH remains above 7.5.
3. Discharge cold through shaking screen filter.

PROPERTIES AND TEST METHODS
PECULIAR TO EMULSION POLYMERS

Emulsion polymers are unique in many of the test methods required to characterise the system. Some of the tests specific to latices will be outlined here.

Freeze Thaw Stability

Latices containing up to 50% water are susceptible to freezing. Upon thawing it is possible for destabilisation of the latex to occur. The normal test method is to cycle the latex between ambient and freezing. A freezer (normally at $-16°C$) is often used overnight to freeze the latex and it is allowed to thaw at laboratory temperature or in a 25°C oven or bath. Stability after 3–5 cycles is considered satisfactory.

Mechanical Stability

Many latices are shear sensitive, particularly to pumping, which may be essential in bulk handling facilities. There are many ways to determine this, but an easy method is to rub the latex between finger and thumb. If it is mechanically unstable it will rapidly 'ball up'. A certain amount of experience is required for this test.

Storage Stability

It is necessary for latex to remain stable upon storage and exposure to light. Storage stability can be tested by examining the condition (i.e. has separation occurred, is there sedimentation, etc.) and determining the viscosity with respect to time. A shelf life of six months is essential. Accelerated tests can be used by storing in ovens at say 40°C. It is important to correlate accelerated tests with normal storage tests. If the latex is to be used in the tropics, then it is essential that storage tests at elevated temperatures are conducted.

Latices should not destabilise by exposure to light. It is possible to test this by leaving jars on the window cill of a south facing window.

Non-volatile Content

The non-volatile content (n.v.c.) or solids as it is commonly referred to is a standard test method in resin chemistry. But there are a multitude of different methods depending upon the volatility of the solvent/water, etc., and thermal stability of the resin. Acrylamide containing resins slowly decompose on heating, and it is generally difficult, if not impossible, to obtain a non-volatile content to constant weight.

A common fault in determining non-volatile content of latices is to use too thick a sample. Not more than 1 g of emulsion should be used in each tray, and in the case of viscous emulsions the sample should be quickly spread with a small spatula before weighing. Drying should be for 30 minutes at 130°C.

The time and temperature can be altered if it is consistantly found that constant weight is not being achieved. Vacuum ovens can be used for thermally sensitive polymers. For some systems it is desirable to move the resin during evaporation of water/solvent and a bent paper-clip, pre-weighed into the 'solids dish', can be used to stir the 'drying film' to avoid a crust forming entrapping liquid underneath. As a general rule, this technique is more important for the faster evaporating organic solvents.

Particle Size

It is important to reproduce the particle size of any latex and also to be able to measure it. Particle sizes generally encountered range from less than one micron for a fine latex, to about three microns for a coarse one.

The distribution as well as the average size are both important. Particle size can be measured by many techniques, which include:

- i) Scanning electron microscope (S.E.M.)
- ii) Ultra centrifuge
- iii) Disc centrifuge
- iv) Light transmission/scattering
- v) Soap titration
- vi) 'Nanosizer'.

Each method gives a different value and can be considered relative. It is necessary to establish standards using one technique. As a general rule, the 'bluer' the latex appears, the finer the particle size for any given polymer system.

Early calibrations were made using the electron microscope which gave reproducible results with polystyrene emulsions. Sample preparation which involves drying and coating the particles with metal to render them opaque to electron beams can cause deformation and coagulation of particles. For the soft polymers often used for surface coatings, electron micrographs can give misleading results. In recent years a new technique which gives a reliable weight average particle size has been developed. Polymer particles undergo Brownian movement; the energy due to thermal effects of any particle depends only on the temperature and a large number of particles have a Gaussian distribution of energies around this characteristic value. When a laser beam is passed into a diluted emulsion, light will be reflected by a particle which acts as a source of light. If the particle is moving towards or away from a detector, the wavelength of the light will change owing to the Doppler principle. By measuring the change in wavelength the particle size can be computed. This method is quick, reproducible and does not involve errors of preparation. This is the principle of operation of the Nanosizer (from Coulter). At one time, disc centrifuges were often used, but the determination could be tedious.

The particle size of an emulsion is an important factor in its application and it is also a useful indicator of stability. A robust recipe will produce latices with a small spread of particle size. There will always be variations in weighing or metering reactants, in reaction temperatures and times as well as raw material variations. If these variations do not cause particle instability, good constancy of particle size will result. The first sign of instability is particle agglomeration, which will give an increase in particle size if moderate and emulsion breakdown if highly unstable.

Viscosity

Viscosity is the resistance to flow in a liquid and is measured in poise. The study of viscosity is known as Rheology and this plays an important part in emulsion applications.

In an ideal liquid, the rate of flow is proportional to the force inducing that flow. Mobile liquids approximate to this ideal behaviour and are known as Newtonian liquids. For most commercial emulsion polymers the rate of flow increases more rapidly than the rise in inducing force. This is known as pseudoplasticity.

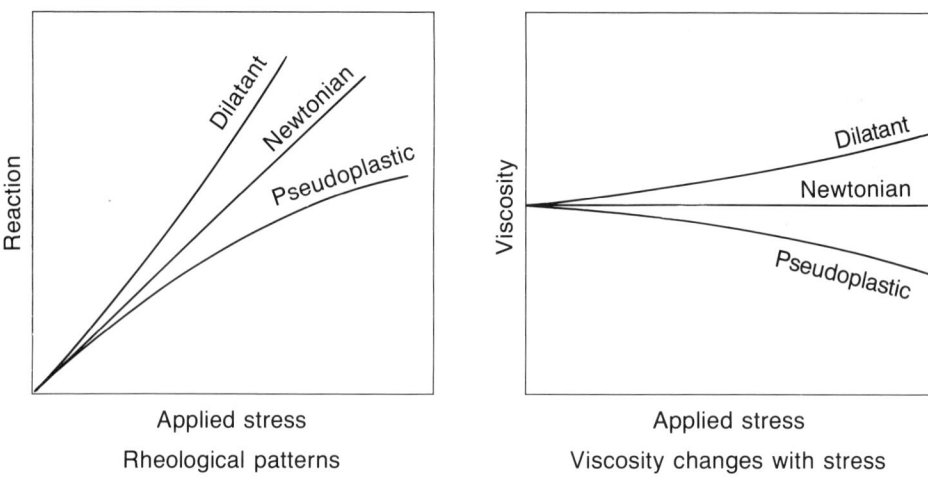

Rheological patterns

Viscosity changes with stress

The term thixotropy has often been confused with the pseudoplastic state which is a reduction in viscosity as the rate of shear increases. Thixotropy is a reduction in viscosity with time. Almost all thixotropic liquids are also pseudoplastic, but not all pseudoplastic liquids are thixotropic. Both pseudoplastic and thixotropic behaviour are caused by 'structure' which involves the interactions between particles in the liquid. The strength of these interactions and the rate at which they form control the rheology of the system. Viscosity measurement is not an instantaneous process and the rate of formation of inter-actions relative to the time of measurement determines the rheology. If interactions are formed relatively fast an equilibrium will be set up between formation (constant) and destruction (rate dependent on shear) of interactions, and the system will appear to be pseudoplastic. If interactions are formed slowly the apparent viscosity will decrease, this is because the viscometer will record a value before equilibrium is reached. In a brushing paint, the removal of brush marks requires a slow rate of forming structure. If the paint is applied to a vertical surface a fast build-up of structure is required to prevent flow of the paint down the surface, a phenomenon termed 'sagging'.

A commonly used viscometer is the Brookfield in which the resistance on a spindle rotating at a fixed speed in the emulsion is recorded. The scale reading is multiplied by a factor dependent on the spindle and its speed. The Brookfield is a cheap and reliable quality control instrument, but it is limited to a small range of shear rates from 5 to 100 rpm, so that it gives little information on the emulsion rheology. The Ferranti-Shirley viscometer measures the reaction on one plate of another (an inverted cone with a small angle making it nearly flat, different cone sizes are available) whose speed of rotation is varied from zero to a large value and then reduced. Typical Ferranti-Shirley plots are shown in the figures below.

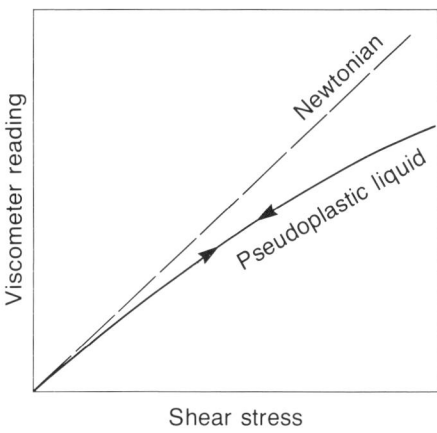

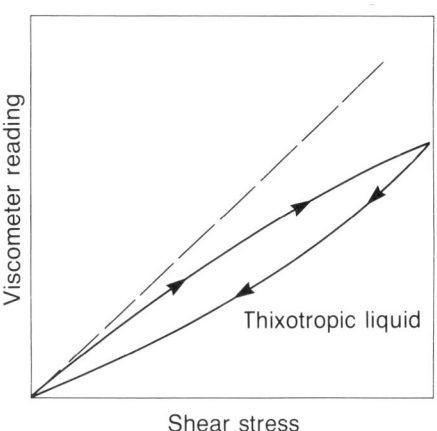

FILM FORMATION

When used as a surface coating, an emulsion must form a coherent film. The evaporation of water during the final stages of drying causes stress in the film and a brittle film will crack under this stress. The temperature at which no stress cracking occurs is termed the minimum film forming temperature, M.F.F.T. or M.F.T. This is measured by a M.F.T. bridge which consists of a stainless steel slab with a slight hollow running along its length which can be filled with emulsion. Each end of the slab is turned down to provide a contact with a thermostatically controlled water bath; three or four small cavities along the length of the slab are used to house thermometers.

If the two bath temperatures have been chosen correctly, the emulsion will dry to continuous film at the hot end of the bridge and will be cracked or crazed at the cold end. The boundary between the cracked and intact films occurs at the minimum film forming temperature. The M.F.T. is generally slightly lower than the Tg because of the plasticising effect of water on the polymer film.

It is critical that the M.F.T. is below the temperature prevailing during the drying of the film, otherwise a discontinuous, hazy film is formed because the polymer did not film form at that temperature.

The formation of a polymer film form or latex film is different to all other types of film formation. Not only has the water to disappear by evaporation or absorption into substrate, but during this process the polymer particles must fuse together.

A polymer emulsion consists of polymer particles suspended in a relatively mobile liquid. The particles are free to move but electrostatic charges prevent particles approaching each other too closely. As the water evaporates, the volume of water separating the particles decreases and the particles are forced closer together. Electrostatic repulsion is reduced by the counter ions, generally sodium, associating with the particle charges. At some stage during the drying, the particles become so tightly packed that they cannot move and a gel structure results. This is the flocculation stage of drying and the gel is dispersible in water. Further drying causes coalescence of the particles which is an irreversible process. The rate of water loss is relatively fast up to the flocculation stage and varies little with particle size. Subsequent drying is much slower with smaller particle sized latices drying more slowly than larger ones.

The latex may be 50% non-volatile content which means that 50% of the initial coating is water. As this is removed the particles are physically drawn together on a simple shrinkage basis. However, a point is reached where the capillary forces of attraction take over and these forces are strong enough to overcome the latex stabilisation forces. The polymer particles touch and for coherent film formation to occur the polymer particles should flow into one another. This is called coalescence. The remaining water should be removed during this stage. Lack of the particles cold flowing gives rise to poor films.

It is often stated that coalescence is caused by surface tension forces, but it is probable that these forces only operate in the final stages of particle coalescence. The pressure needed to cause particle deformation prior to coalescence may be caused by film contraction caused by loss of water. Soft films deform more readily than harder ones and this means that soft polymers require less energy to start coalescence. When coalescence starts, soft polymers flow more easily and therefore integrate more readily into the film.

In an emulsion the interfacial tension between the polymer and water is reduced by surfactant at the interface. Once the water has evaporated the surfactant will tend to form inverted micelles with the hydrophilic groups forming the core. Sulphate end groups and any grafted surfactant or colloid on the particle surface will provide a nucleus for such inverted micelles. Most colloids and some ionic surfactants are not compatible with polymers and will form a dispersed phase giving a cloudy film. Most nonionic surfactants and polyvinyl pyrrolidone are soluble in polymers and would not reduce film clarity. It is often claimed that large particle size emulsions give films with a lower tensile strength than that for a smaller particle emulsion of the same polymer. The argument runs that large particles do not fully integrate during drying of the water and that the discontinuity in the film reduces the film strength. The gloss of an unpigmented film from an emulsion polymer increases as the particle size decreases. This finding has led to a range of small particle size polymer emulsions being suggested for gloss or semi-gloss applications.

Like the theory of emulsion polymerisation there are many differing theories about the mechanism of film formation. The reader is advised to consult specialist works for more details.

The common themes from many of the theories relate to the following 'driving forces' for coalescence.

  i) Capillary forces.

  ii) Surface tension of both polymer/water and water/air interfaces.

  iii) Inter-diffusion of both polymer and surfactant molecules between contiguous particles.

  iv) The resistance of any particle to physical deformation (i.e. 'cold flow').

The improvement to film properties with respect to time is believed to be evidence for autohesion (inter-diffusion) where residual levels of water and solvent do not vary.

Water can be considered to be removed in three stages. Vanderhoff et al (J. Polymer Science Symposium, No: 41, 155 (1973)) defined the three stages as; i) An initial constant evaporation stage (or absorption) where the particles still retain some mobility; ii) An intermediate stage, when particles start to come into irreversible contact and the rate of water evaporation is much less than that in stage (i) (some 5−10% of the rate); iii) Loss of water from the film by diffusion which is very slow and can occur over days. It is believed that the water diffuses along hydrophillic networks.

Loss of water by evaporation alone means that water at the surface of the latex is lost first creating a concentration gradient between the top and bottom of the film. Coalescence occurs first on the surface and further loss of water is controlled by the rate of diffusion (evaporation rate is higher than diffusion rate).

When the substrate absorbs water, water is lost from the base of the film until the substrate reaches saturation point, Obviously the absorbency of the substrate is important to the mechanism of the film formation. If the substrate is too absorbent then all the water is removed before coalescence can commence. An everyday example to overcome this is the excessive dilution of a latex paint for the first coat on a very porous substrate.

Thus, the initial removal of water from the film depends upon the balance of loss by evaporation and absorption.

It is important to realise that at some point during film formation an irreversible change must occur for the coating to be of any practical use.

Coalescing solvents are believed to work in two ways. Sullivan (J. Paint Tech. 47 (610), p. 61, (1975)) considered two types of coalescing solvents and their different mechanisms. The film is considered to have a hydrophillic network where polar coalescing solvents like ethylene glycol partition, and a polymer phase where solvents like ethylene glycol monobutyl ether acetate partition.

Ethylene glycol facilitates coalescent evaporation because it swells the hydrophillic network, thereby creating a larger pathway for diffusion. The rate of initial water evaporation is unaffected by coalescing solvents, but they may retard final evaporation. The more polar the solvent the faster it will evaporate. The use of a water soluble solvent enables water evaporation to be faster than the rate of diffusion during the final stages. In the absence of solvents, or in the presence of water insoluble solvents, water evaporation is diffusion controlled.

Thus, most practical latex systems require small additions of coalescing solvents to increase the drying rate. These solvents are generally not retained in the film despite having low evaporation rates. It is possible to mix the solvent so that one partitions to the hydrophillic network and the other to the polymer phase. Surprisingly they do not interfere with the action of each other.

Obviously the particle size is important because it affects the capillary and other film formation forces. However, varying particle size has minimal effect on M.F.F.T. The Tg of the comonomers and the presence of external plasticisers are more important for film formation.

Colloids have an important effect upon film formation. Achievement of physical properties of a film occurs more rapidly in the presence of colloids, with colloidal particles bridging gaps between polymer particles. Indeed colloidal particles can fill residual voids in the resin matrix, thereby increasing the strength of the film. Colloid free systems can take days before optimum strength is attained.

When porous substrates are to be coated it is important that the particle size of the latex is larger than the size of the pores, to stop the particles being absorbed into the pores. It is in this area that colloids can have an important effect on film formation. Colloid containing latices tend to have a larger particle size than a colloid free system. In addition the colloids can be large enough to stop absorption into the pores by bridging the pores.

The pigment volume concentration influences solvent evaporation. Below the critical pigment volume concentration (C.P.V.C.) non-porous pigments act as barriers to solvent passage, whilst above the C.P.V.C. voids, and polymer discontinuity affect solvent loss. At concentrations around the C.P.V.C. solvent evaporation rates are minimal.

Pigmentation of Latices

The pigmentation of latices is different from that of most resin systems. In most systems the resin (binder) is soluble in the liquid system. Pigment is dispersed in the solution which can be considered as a varnish (for inks). Resin can be absorbed onto the pigment. Partially soluble resins can coat the pigment. But there is a significant amount of soluble resin present in the pigmented system, and evaporation (or loss) of solvent results in the resin being deposited over the surface in a layer which may vary in thickness, depending upon physical forces and inter-reactions during drying.

In a latex system there is essentially no soluble binder resin. The pigment particles must be dispersed with polymer particles. The polymer particles and pigment particles must stick together. Too much sticking and flocculation occurs. Pigment particles can induce charge destabilisation, particularly some of the modern pigments which are surface treated.

Thus, a pigmented latex consists of agglomerates of pigment and polymer particles. However, the pigment acts as a non-deformable particle, similar to a monomer of high Tg. Thus, during film formation the polymer particles must be capable of deforming more than in a non-pigmented system. This also requires the loss of water to be slow enough to allow this to happen, otherwise incomplete film formation occurs. Not only must the polymer particles flow into each other, they must flow around intervening pigment particles.

Obviously the nature, surface and surface area of the pigment particles have important effects on latex stability and subsequent film formation. However, for any given pigment, the polymer particle size, Tg and surfactant level of the latex are critical. Determination of the C.P.V.C. must take into account the particle size of the polymer as well as the pigment void volumes.

One of the important properties is for the pigment not to flocculate. The smaller the polymer particles the higher the C.P.V.C., and this also applies for softer particles which can deform and stick more readily.

The reader is recommended to consult a paint technology text book for more detailed descriptions of the effects of pigment concentrations, below and above, the C.P.V.C. on the performance of the film.

APPLICATION OF EMULSION POLYMERS IN PAINTS

The original water based paints, distempers, were prepared by emulsifying an oil bound paint in water. They had a limited use on porous surfaces where oil absorption into the substrate was reduced in relation to oleoresinous finishes. The replacement of distempers by pigmenting large particle size polyvinyl acetate emulsions, gave faster film drying and better 'breathing'. Most of the early problems derived from the dispersed nature of the polymer. In a conventional oleoresinous paint the resin contributed directly to the vehicle viscosity and to dispersion of the pigment, whereas in a latex paint the polymer particles acted more like pigment than vehicle. It was found that additions of colloid, sequesterant and humectants were necessary to give good performance. Latex paints have lower pigment volume contents (P.V.C.) than oil paints. This is partly because high P.V.C. adversely affects film formation; in oil based undercoats and matt finishes, a high P.V.C. is necessary to prevent gloss while small quantities of pigment will produce mattness in latex paints. The three types of additives will be discussed separately.

Humectants

When a latex paint is applied whether by brush or roller, the rate of drying must be reduced to allow time to brush out the paint.

High boiling water soluble liquids are added to latices to reduce the rate of evaporation of water. Ethylene and propylene glycols are commonly used at $5-10\%$ of the latex weight for vinyl acetate emulsions and up to 20% for acrylics and styrene-acrylics. Neither glycol interferes with the particle coalescence process. Hexylene glycol is sometimes used at $5-10\%$ on latex weight and this retards the water evaporation in the early stages of drying; in the latter stages it can dissolve in the polymer particles, especially in the case of vinyl acetate copolymers. When dissolved in the polymer, hexylene glycol reduces the Tg of the polymer, making it softer and easier to deform prior to coalescence. Butyl carbitol is slightly soluble in water, although less so than hexylene glycol, also it reduces the rate of evaporation only slightly, but it is an efficient coalescent solvent. Butyl cellosolve acetate is insoluble in water and acts only as a coalescent.

Pigments

Most of the common pigments and extenders can be used with emulsion polymers. The main problem is that pigments can adsorb surfactant from polymer particles and de-stabilise the latex. Calgon is often used as a dispersant and a surfactant or benzene sulphonic acid may also be added. These additives ensure thorough wetting of pigments and prevent destabilisation of the latex. A colloid is also added as an aqueous solution to assist pigment dispersion and also to increase the viscosity. Hydroxyethyl cellulose is a commonly used colloid for this purpose, but methyl cellulose, ammonium salts of poly(acrylic acid) and poly(vinyl alcohol) have also been used. Both vinyl acetate and acrylic ester polymers have refractive indices below 1.55 which enhances the opacity of pigments used with them.

Other Additives

Anti-foaming agents, preservatives and stabilisers can all be added to latex paints. Polymeric de-foamers of the Bevaloid type are popular, as they combine long life and do not cause wetting or livering troubles. Silicone de-foamers can cause cissing and other surface faults, while the nonanol type de-foamers tend to dissolve in the polymer particles and are then ineffective.

Vinyl polymers are not normally susceptible to bacterial or fungal attack, but many colloids, especially those based on cellulose, and surfactants can be bio-degraded. Residual monomer is quite an effective biocide but permanent preservatives are required to prevent degradation of the dried film. Esters of p-hydroxybenzoic are popular, but a wide range of proprietary products are also employed. The use of mercury containing biocides is discouraged on environmental grounds. Formalin type biocides were popular before fears about the toxicity of formaldehyde. Formaldehyde must not be used in any form with vinyl chloride containing latices, because a highly toxic gas similar to phosgene is formed.

Vinyl chloride and vinyl acetate polymers are discoloured when exposed to heat or light. Polymers containing long sequencies of either monomer should have some calcium carbonate as an alkaline extender and an organotin stabiliser. Styrene copolymers can discolour in bright light and a U.V. stabiliser can be added to suppress this reaction.

The M.F.T. of the emulsion as formulated must be lower than any temperature at which the paint is likely to be applied. For an indoor paint in the UK it is unlikely that temperatures below 10°C will be encountered while the outdoor limit might be 5°C. Damage to paint films can occur should the ambient temperature fall below the Tg of the film. The Tg of a hard polymer, such as vinyl acetate, styrene or methyl methacrylate, can be reduced either by copolymerisation with a soft monomer or by adding an external plasticiser, such as di-butyl phthalate. On a cost effective basis external plasticisation is better for reducing the Tg. The snags are, firstly, that plasticisers are slowly lost through evaporation and secondly they can also diffuse into the substrate. The use of an internally plasticised polymer overcomes both problems and also saves the processing time necessary to incorporate an external plasticiser.

Ethylene, Veova and 2-ethylhexyl acrylate are most often used to soften vinyl acetate. Ethylene is the cheapest of the three monomers and is also the most efficient softener (see the table below). It requires a very large capital investment for a high pressure reactor and the relatively long reaction times are expensive. 2-ethylhexyl acrylate is difficult to incorporate because of adverse reactivity ratios and it also gives the weakest films of the three copolymers. Veova is the easiest of the three monomers to incorporate and it gives films with surprisingly good water and alkali resistance. It is not an efficient softening monomer on a weight basis and very inefficient on a cost basis. The use of alpha-olefins such as 1-hexene or 1-dodecene is popular in patents, but the olefins are very difficult to polymerise by a free radical mechanism and they are not used commercially.

Tg FOR SOME VINYL ACETATE COPOLYMERS TO ILLUSTRATE THE EFFECTIVENESS OF DIFFERENT PLASTICISING COMONOMERS
(Calculated in K)

Wt % comonomer	5	10	15	20	25
Comonomer					
Ethylene	266	240	220	209	198
2-ethylhexyl acrylate	301	297	294	290	286
Veova	303	301	300	298	296

Acrylic ester polymers and their copolymers with styrene or methyl methacrylate can be readily prepared with a wide range of Tg, but vinylidine chloride presents problems. In the table on page 163 its Tg is given as 255 K, which is the glass transition temperature of the amorphous polymer. Polyvinylidine chloride behaves as a crystalline polymer for which the glass transition occurs at about 370–410 K and calculations of glass transition temperatures are pointless for this monomer.

An important property of polymer films is their permeability to gases and vapours. Vinylidine chloride and its copolymers have exceptionally low permeability, while poly(vinyl acetate) has the highest permeability of commercial polymers. Acrylate and styrene acrylate polymers have intermediate values with the presence of styrene increasing diffusion of gases. For a specific polymer, permeability will vary with the nature of the gas or vapour and permeability will also increase with an increase in pigmentation. The transport of water vapour through polymer films is important both in the drying process and in use. Rapid transport will tend to occur when the polymer is relatively hydrophillic, e.g. poly(vinyl acetate), or contains acid or amide groups, while it will be much slower for hydrophobic polymers.

Paints in bathrooms and kitchens will require high permeability to water vapour to avoid condensation of water. Coatings on wood and metal should have low water permeability to reduce decay and rusting. Other applications have their special require-ments for permeability. The inherent permeability of polymer films represents the minimum obtainable with a coherent film. When low vapour transmissions are required it is important to ensure that the film produced from the latex is as integral as possible and extra coalescent solvent may be added. Should high vapour permeation be needed the formation of minor discontinuities in the film may be encouraged by not adding any coalescent.

Formulations for Coatings and Inks based upon Emulsion and Aqueous Vinyl and Acrylic Polymers

Both the emulsion polymer and paint industry guard their industrial secrets. Polymer manufacturers will state the type of polymer and of surfactant or colloid, but will not divulge quantities of either monomer ratios or of surfactants concentrations. They will say whether surfactants are anionic or nonionic, but will not divulge the chemical nature of individual surfactants used, nor of grades or colloids. Polymer and pigment suppliers will provide recipes showing how their products may be used in paints and these recipes do work. Paint manufacturers have their own secret additives and their concentrations which they feel give better performance than the general recommendations of their suppliers. In the examples given below, the recipes are taken from suppliers literature, and are selected to illustrate general points of formulation rather than to provide instant paints.

Paints utilising Thermoplastic Emulsions

MATT WHITE INTERIOR PAINT

Titanium di-oxide	19.020
Micro-talc A.T.1	7.390
Microdol 1	22.180
Calgon S (4% in water)	2.360
Manoxol OT (12% in water)	0.197
Cellosize QP 4400 (3% in water)	12.910
Bevaloid 677	0.059
Preventol CMK (10% in water/methanol)	0.690
Vinamul N 6810 (55% non-volatile content)	22.960
Vinamul N 6825 (55% non-volatile content)	7.690
Ethylene glycol	0.986
Water	3.558
	100.000

Properties

p.v.c. 50%

Mixing and dispersion of these ingredients would have been by ball milling in the early years of emulsion paint manufacture. Modern practise is to roughly mix the ingredients and to pass them through a high speed mixer or colloid mill.

The Vinamul emulsions are vinyl acetate-acrylate copolymers with different amounts of acrylate. The reason for using a mixture of two copolymers may be to obtain a softness intermediate between the two grades. But it could also be to obtain improved film integrity at a given hardness utilising the easier deformation of the softer copolymer. The p.v.c. of 50% is lower than an oleoresinous paint of the same opactity which would be about 70%. To compensate for the lower pigment loading a thicker film might be applied, but opacity can be increased if necessary by increasing the ratio of titanium di-oxide to talc.

SEMI-GLOSS ALL ACRYLIC PAINT

25% sodium benzene sulphonate	0.669
Balab 748	0.191
Propylene glycol	4.870
Rutile titanium di-oxide	22.540
Grind these components for 20 minutes on a disc disperser	
Rhoplex AC-22 (44.5% non-volatile content)	49.948
Rhoplex AC-73 (46% non-volatile content)	10.700
Butyl cellosolve	0.573
Acrysol G-110 (22% non-volatile content)	0.478
Water	9.840
Triton GR-5 (60% non-volatile content)	0.191
	100.000

Properties

Non-volatile content 50%

p.v.c. 19%

The polymer emulsions had M.F.T.'s of 8°C and 35°C respectively. In this example, a mixture of hard and soft polymer latices are used deliberately as the manufacturers produce an all acrylic polymer with intermediate hardness. A small amount of coalescing solvent is added to further improve film integrity. The use of a much larger amount of glycol compared with the previous example is common when smaller particle size emulsions are used. Propylene glycol is preferred to ethylene glycol for acrylic polymer latices because of its lower toxicity; its use with vinyl acetate polymers is less popular because of possible absorption of propylene glycol into polymer particles. Because latices give poor gloss films compared with solution polymers, gloss and semi-gloss latex paints often do not contain extenders. The lower cost of polymer emulsions relative to solutions helps to offset the increase in cost. A final point is the use of an acrylic thickener rather than the cellulosic colloids preferred for vinyl acetate finishes.

WASH RESISTANT INTERIOR MATT PAINT

Titanium di-oxide, Runa RE376	20.79
Micromica W160	5.70
Whiting BWF42	11.79
Calgon S (5% aqueous)	2.90
Celacose CP20 (2% aqueous)	20.09
Butyl carbitol acetate	0.70
Di-butyl phthalate	0.70
Foamaster NXZ	0.20
Acticide MPM	0.04
Water	12.10
Vinamul 3401	24.99
	100.00

Properties

pH 7.6

Non-volatile content 51%

Viscosity 610 cps at 25°C

p.v.c. 52%

Vinamul is a vinyl acetate-vinyl chloride-ethylene terpolymer with an M.F.T. of 14°C. Some coalescing solvent is added to assist film formation that is required for this application. No addition of glycol is required as the open time of large particle size emulsions is fairly long without added humectant. Note also that a cellulosic thickener is preferred with a vinyl acetate copolymer.

SILK LUSTRE FINISH USING
VA-ETHYLENE COPOLYMER

Titanium di-oxide, Tiona 535	29.440
Microdol extra	2.790
Nopco NXZ	0.200
Dispex G40	0.599
Natrosol MR, 2% aqueous	7.480
Water	2.290
Mix and grind, then add:	
VA-E emulsion polymer, 50% solids	41.431
Water	4.490
Natrosol MR, 2% aqueous	7.480
Propylene glycol	2.000
Butyl carbitol acetate	1.700
Nuodex 321E	0.100
	100.000

Properties

Non-volatile content 57.5%

Viscosity 90 cps at 25°C

p.v.c. 32%

A reduction of 20% in the p.v.c. from 50 to 30% is sufficient to impart enough gloss to give a silk finish. Note that the silk lustre has virtually no extender.

EXTERIOR ACRYLIC PAINT

Hydroxyethyl cellulose (2½%)	8.370
Water	5.860
Tamol 731 (25%) (dispersant)	1.260
Triton CF-10	0.209
Anti-foam	0.084
Ethylene glycol	2.090
Rutile titanium di-oxide	20.930
Talc	17.580

Mix and grind in a high speed mill then add:

Rhoplex AC-388 (50% non-volatile content)	38.511
Anti-foam	0.084
Ammonia solution (28%)	0.167
Tri-butyl phosphate	0.921
Propylene glycol	2.930
Preservative	0.167
Water	0.837
	100.000

Properties
pH 9.5
Non-volatile content 58%
p.v.c. 40%

The technique of grinding a pigment paste before adding the polymer emulsion is often used. There is a small economic gain in milling time but the main advantage is that the latex is not subject to shear. The high shear inherent in milling can cause coagulation of particles in the emulsion; such coalescence can clog the rotor on the mixing mill. The latex is a carboxylated acrylic copolymer where the acid groups have been ionised by raising the pH to 9–10; the presence of such ionised groups in the polymer particles confers freeze-thaw stability to latices. The low M.F.T. of 7°C gives the film good integrity required for an exterior finish with a minimum of coalescing solvent.

POLY(VINYLIDINE CHLORIDE) GLOSS PAINT

Texicryl 13-011	8.190
Add with stirring a mixture of:	
Hexylene glycol	2.710
Ammonia solution (35%)	0.266
To the clear solution add:	
Titanium di-oxide	20.160
Disperse using a high speed impeller mill and add:	
Propylene glycol	10.230
Water	2.660
Add a premix of:	
Polidene 33-061	54.249
Texicryl 13-011	0.460
Texicryl 13-301	0.256
Ammonia solution, 0.880	0.819
	100.000

Properties

Non-volatile content 53%

p.v.c. 16.5%

The two Texicryl emulsions contain acrylic esters copolymerised with 5−10% acrylic acid which become water soluble when made alkaline. They behave both as thickeners and as wetting agents. In the present application they have the additional advantage that they are compatible with the vinylidine chloride-acrylic ester copolymer. Colloids which are incompatible would reduce the gloss. The small particle size, 0.18 micron, of the Polidene emulsion requires a large addition of glycol to give good open time. Good film integrity and very low film permeability endows this finish with excellent weathering characteristics.

ACRYLIC FLOOR PAINT

Tamol 731 (25% aqueous)	0.888
Triton CF-10	0.178
Ethylene glycol	2.660
Water	6.570
Rutile titanium di-oxide	11.460
Silica	11.100
Chromium oxide X-1134	6.220
Grind on a Cowles dissolver at 6000 rpm for 10 minutes and add:	
Rhoplex AC-61	49.469
Di-actone alcohol	2.220
Butyl cellosolve	2.220
Colloid 600 (anti-foam)	0.355
Hydroxyethyl cellulose WP-4400 (3%)	5.330
Aluminium oxide	1.330
	100.000

The Tg of the Rhoplex AC-61 is 18°C which is close to the ambient temperatures to which the finish will be subjected. The hardness and toughness of a polymer film reach a maximum as the film is cooled to its glass transition and film integrity is enhanced by the use of coalescents. The paint is designed for concrete floors and the pigments contribute to performance. The rutile provides covering power and the chromium oxide a dull green colour which avoids the glare found with white finishes. Silica is an extremely hard material and reinforces the film. The aluminium oxide gives a large increase in abrasion resistance without reducing the gloss.

BLACK ACRYLIC EXTERIOR PAINT

Water	2.030
Tamol 731 (25% non-volatile content)	0.582
Triton CF-10	0.185
Anti-foam	0.185
Ethylene glycol	2.310
Hydroxyethyl cellulose (2.5%)	12.920
Lamp black	1.850
Silica	31.570
Pine oil	0.277
Grind for 12 minutes and then add with stirring:	
Rhoplex AC-35	47.260
Preservative	0.646
Ammonia solution (28%)	0.185
	100.000

Properties

pH 9.5

Non-volatile content 55%

p.v.c. 40%

A carboxylated acrylic copolymer has a higher water permeability than an unmodified acrylic and this could detract from its performance on wood. The extremely good resistance of acrylics to heat and light gives good durability. A black gloss paint tends to heat up in direct sunlight which can cause deterioration to many film formers. Acrylic finishes are particularly suitable for dark coloured exterior paints.

ROOF COATING BASED ON
THERMOPLASTIC EMULSION

Styrene-acrylic emulsion (65%)	70.0
China clay	10.0
Titanium di-oxide	7.5
Durcal D5	6.3
AT1 talc	3.0
Ammonia	0.2
Anti-foam	0.5
Preservative	0.5
Butyl carbitol acetate	1.0
White spirit	1.0
	100.0

DECORATIVE EMULSION PAINT
BASED ON TERPOLYMER LATEX

Terpolymer (as solid)[1]	8.5
Water	25.0
Titanium di-oxide	12.0
Whiting BWF42	30.0
Cellulosic thickener	22.0
Z – (Zl butoxyethoxy) – ethyl acetate	2.0
Sodium hexametaphosphate	0.1
Potassium hydroxide	0.1
Anti-foam	0.2
Preservative	0.1
	100.0

(1) Pressure terpolymer of vinyl acetate/vinyl chloride/ethylene.

WHITE EMULSION PAINT BASED ON
VINYL ACETATE HOMOPOLYMER

Titanium di-oxide	11.79
Talc	5.89
Chalk	5.89
Lithopone (30% ZnS)	15.72
Methylcellulose (4%)	5.89
Sodium hexametaphosphate (5%)	2.36
Water	9.82
Ammonia (25%) until pH = 8.0	
Grind and add:	
Vinyl acetate homopolymer (60%)	23.58
Plasticiser	3.14
Methylcellulose (4%)	3.93
Water	11.79
Fungicide	0.20
Ammonia (25%) until pH = 8.0	100.00

VINYL ACETATE –
VINYL VERSATATE EMULSION PAINT

Copolymer emulsion (55%)[1]	25.430
Titanium di-oxide	15.430
Whiting	17.250
Talc	3.540
Ethyl hydroxyethyl cellulose (4%)	14.070
Potassium polymethacrylate (5%)	1.270
Calgon 5 (5%)	1.270
Potassium hydroxide (50%)	0.182
Butyl di-glycol acetate	1.820
Anti-foam	0.182
Fungicide	0.036
Water	19.520
	100.00

(1) A 90:10 copolymer of vinyl acetate:vinyl versatate.

Properties

p.v.c. 60%

Non-volatile content 50%

VINYL ACETATE –
ACRYLIC EMULSION PAINT

Vinyl acetate/2-ethylhexyl acrylate copolymer emulsion (55%)	30.00
Titanium di-oxide	27.00
Barytes	8.00
Whiting	3.50
Ethyl hydroxyethyl cellulose (5%)	4.00
Calgon 5 (5%)	1.00
Fungicide	0.05
Anti-foam	0.10
Butyl di-glycol acetate	2.00
Water	24.35
	100.00

Properties

p.v.c. 40%

Non-volatile content 55%

Pigment/binder ratio 2.25/1

Paints utilising Thermosetting Emulsions

WATER BASED ACRYLIC STOVING ENAMEL

Mill together:

Cymel 303	5.880
Tamol 731	0.466
Butyl cellosolve	1.490
Water	9.790
Di-methylaminoethanol (DMAE)	0.093
Rutile titanium di-oxide	19.590

Add with stirring:

Acrysol WS-68	60.641
Di-methylaminoethanol (15% aqueous)	1.490
p-toluene sulphonic acid (10% aqueous)	0.560
	100.000

Properties

p.v.c. 40%

The finish is applied by spraying
and stoved at 150°C for 30 mins.

Acrysol is a small particle size emulsion of an acrylic copolymer containing acid groups which solubilise the polymer when the amine is added. Cymel 303 is 'monomeric' melamine-formaldehyde resin which cross-links the acid groups in the acrylic copolymer. The p-toluene sulphonic acid catalyses the cross-linking reaction after the D.M.A.E. evaporates during the stoving operation.

This example is one of three which indicates a variety of approaches possible for the important field of industrial stoving finishes, for which alkyd, urethane, epoxy, polyesters and cellulosic formulations are also used. In comparing the three examples (two from solution polymer section – vinyl chloride stoving enamel and acrylic stoving enamel), it must be stressed that there is no best product. They are different and have both advantages and disadvantages which may vary according to the end use. The water based acrylic is the cheapest of the three finishes, but it can give poor adhesion because of wetting problems. The solvent based acrylic gives good all round properties, but both the cross-linked enamels liberate formaldehyde on curing for which there is a limit of 1 ppm in air. Some organisations will not allow any formaldehyde based products on their sites. The vinylite recipe is the most expensive because of the relatively large solvent content, but it has no undesirable properties, and of the three examples, it is the easiest to recoat.

Water soluble melamine resins are used to cure aqueous alkyds, acrylic, epoxys and polyester resins, by exactly the same chemistry that is discussed earlier in this chapter.

Fully methylated hexamethylol melamine, which is extensively used in water-based coating, has the following structure:

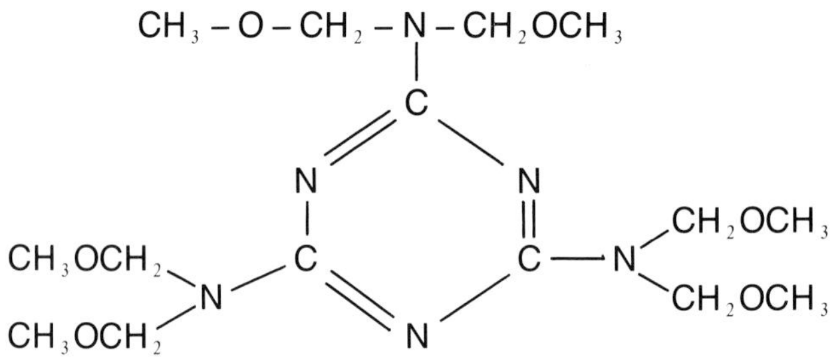

$$CH_3-O-CH_2-N-CH_2OCH_3$$

$$CH_3OCH_2 \diagdown N-C \diagup N \diagdown C-N \diagup CH_2OCH_3$$

$$CH_3OCH_2 \diagup \qquad \diagdown N \qquad \diagdown CH_2OCH_3$$

MHMM

WHITE GLOSS STOVING ENAMEL BASED ON AQUEOUS ACRYLIC RESIN

Methylated melamine formaldehyde resin (88%)	5.0
Titanium di-oxide	24.0
Di-propylene glycol	3.0
Di-methylamino methylpropanol	2.0
Water	7.0
Grind, then add:	
Acrylic resin (45% in water)	55.0
Levelling agent	0.1
Thickener	3.9
	100.0

Cure for 30 minutes at 150°C.

CLEAR FLAT FURNITURE FINISH

Part A	
Flatting agent	2.5
Ethanol	3.0
Ethylene glycol	3.0
Water	7.0
Part B	
Urea-formaldehyde resin	20.0
Dipentine	0.5
Part C	
Hydroxyacrylic emulsion (40% solids)	49.7
Water	14.0
Silicone oil	0.3
Di-methyl ethanolamine to pH = 7.5	100.0

PROCEDURE
1. First mix separately parts A, B and C.
2. Mix part A with part B.
3. Add this mixture to part C and stir, and filter.
4. Immediately before this varnish is used, para-toluenesulphonic acid is added till the pH is less than two. This gives a pot life of approximately eight hours.

MATT FINISH FOR FURNITURE

Part A	
Flatting agent	2.0
Ethanol	2.5
Ethyl glycol	3.0
Allow 15 minutes to swell.	
Part B	
Water	6.0
Melamine resin	21.0
Pine oil	0.4
Tri-butyl phosphate	0.5
Slip agent	0.5
De-foamer	0.1
Slowly stir A into B.	

Part C

Hydroxy acrylate emulsion	50.0
Silicone oil (50% in water)	0.5
Di-methylethanolamine (to pH = 7.5)	0.2
Water	11.8
Sodium hexametaphosphate (10% in water)	1.5
	100.0

Add A/B mixture to C, filter and add immediately prior to use para-toluene sulphonic acid solution to pH = 1.5.

WHITE, COIL COATING FORMULATION

Hexamethoxymethylol melamine (98%)	3.8
Butyl carbitol (or coalescing agent)	1.1
Water	0.6
Butyl glycol	5.4
Di-methylethanolamine	0.3
Rutile titanium di-oxide	30.4
Grind, then add:	
Acrylic dispersion (50%)	53.2
Polypropylene glycol	1.1
Polyethylene wax dispersion	3.4
Ethylene glycol	0.7
	100.0

Stove in 260°C oven with pack metal temperature of 230°C.

Properties

p.v.c. 20%

Non-volatile content 63%

PRINTING INKS

As already stated, emulsion polymers are little used in printing inks. Aqueous solutions are used in flexographic printing, particularly as swelling of the stereos, which can occur with solvent systems, is irradicated.

Many flexographic aqueous systems are used for paper and cardboard packaging inks, and many in the USA utilise emulsion or emulsified solution resins. Paper or board is an ideal substitute for water based packaging inks. The water 'wicks' in very quickly giving a rapid touch dry ink. In some applications the end product is not seen by the consumer, therefore lower quality for economic reasons can be accepted. An example would be a cardboard box containing confectionery packets. The quality of the printing on the individual packets is normally of a much higher standard than the disposable box they were supplied to the shop in.

An example of a gravure application using an emulsion polymer is given before flexographic ones.

Gravure Inks

PRINT BONDING OF NON-WOVEN FABRICS

Emulsion	1.00
Water	1.00
Phthalocyanine blue	0.20
Incorporate using a high speed mixer and add:	
Emulsion	24.95
Water	72.85
	100.00

The polymer emulsion is prepared by methods already outlined, where the monomer addition is made in an emulsion. The copolymer composition is 70% butyl acrylate, 27% methyl methacrylate and 3% N-methylolacrylamide. The emulsion has a particle size of 0.25 micron and contains 45% non-volatile content.

The normal process for producing non-woven fabrics involves saturation of staple fibres with a polymer emulsion, which gives a fabric on drying where the fibres are bonded by acrylic polymer.

When highly absorbent properties are required from the non-woven fabric the bonding is applied on 25−50% of the surface area by the print bonding process.

This is a gravure process using cells from 0.5−1 mm deep; the combination of low emulsion solids and low roller speeds (compared with conventional printing machinery) prevents drying on the cylinders. On the printed area about 40% of binder on fibre weight is applied giving an overall pick-up of up to 20%. The print pattern is a regular arrangement of square or circular shapes. N-methylolacrylamide is prepared by reaction of acrylamide with formaldehyde:

$$CH_2 = CHCONH_2 + CH_2O \rightarrow CH_2 = CHCONHCH_2OH$$

Acrylamide   N-methylolacrylamide

Two methylol groups react, especially with acid catalysts, to give a methylene bridge between the two acrylamide molecules:

$$2\ RNHCH_2OH \rightarrow RNHCH_2NHR + CH_2O$$

$$\text{where R is } CH_2 = CHCO$$

This is the most common method of cross-linking of acrylic and vinyl acetate copolymers. In this application cross-linking of the polymer improves the strength of the non-woven fabric in the presence of water or solvent.

The composite nature of the product effectively gives a fabric less flexible than the binder polymer. For this reason a very soft polymer is used and the Tg in this example is $-14°C$. A cross-linking vinyl acetate, butyl acrylate of the same Tg would give a fabric with much less strength in the presence of water or solvent. A cotton fibre would be used in this application because of its high absorbency of water compared with synthetic fibres.

Flexographic Inks

The flexographic process was originally derived from rotary letterpress printing when dye solutions were applied. The use of low viscosity solutions required minor modifications to the ink application rollers and also needed some method for removing solvent. The process uses stereos made from natural or synthetic rubber and any solvents used in the inks must not swell the stereo. This effectively limits the solvent to water and the lower alcohols, although small amounts of low boiling ketones or esters can be tolerated. The most important polymer type used in flexographic inks is the water soluble acrylic copolymer prepared as an emulsion and solubilised by ammonia into an aqueous solution. Such solutions are termed 'colloidal' in most supplier's literature. The molecular weight of a polymer of this type is typically between 5000 and 30,000 which is rather low for a colloidal particle. It is probable that several molecules associate to form structures similar to the micelles formed by surfactants. A micelle of sodium stearate is believed to comprise $50-100$ molecules of the soap. The same number of ionic groups would be given by $4-8$ molecules of a copolymer of molecular weight 10,000 containing 10% by weight of acrylic acid. The viscosity of the polymer varies with the pH as shown in the following figure.

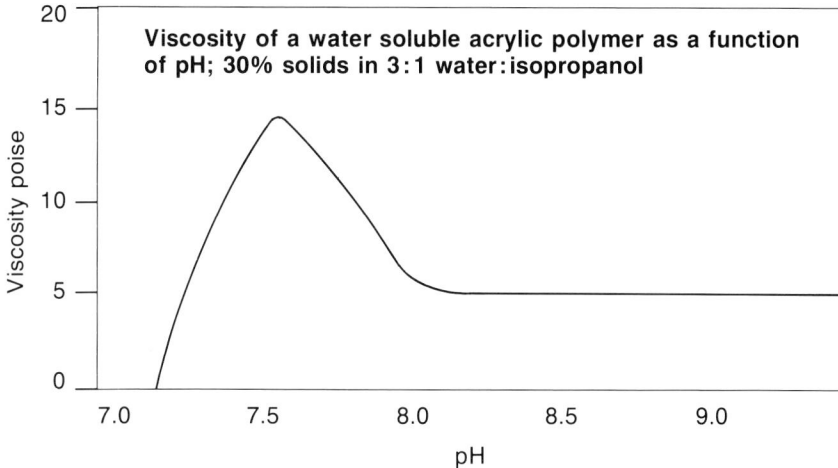

The polymer is insoluble below a pH of 7 and remains in emulsion with a low viscosity. A maximum viscosity occurs at pH 7.5 which indicates that the individual molecules have an extended form caused by the repulsion of the anions. On adding further ammonia, the 'micelle' structure is formed and the repulsion of the anions is reduced by the presence of ammonium gegenions. At lower solids, the viscosity maximum may occur at slightly higher pH (7.6–7.7). The effect of solvent composition on viscosity is shown in the figure below. The explanation of the reason why a relatively small amount of isopropanol should have such a marked effect on the shape of the curve is not easy to understand, but depends on the alcohol affecting the double layer of electrical charge.

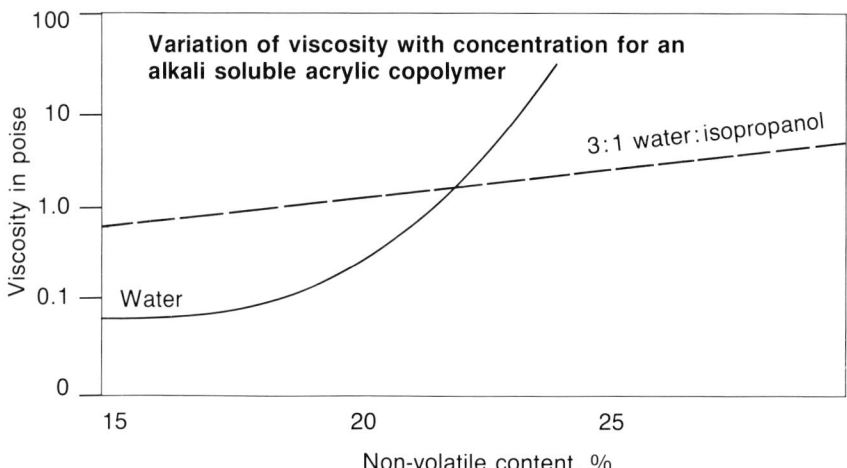

The very sharp increase of viscosity for aqueous solutions above about 22% non-volatile content can cause problems during printing. The addition of an alcohol, normally 5−30% of the weight of water used, allows higher polymer solids to be achieved without viscosity increase in the ducts. Water based stoving finishes often use di-methyl-aminoethanol instead of ammonia to neutralise acid groups. The use of a tertiary amine precludes the possibility of amide formation and gives improved viscosity retention during storage. The stoving removes the amine by evaporation. In many flexographic inks, the use of another tertiary amine is common. Tri-ethanolamine gives the required storage stability and also does not evaporate from the ink duct or rollers during printing.

BLUE FLEXOGRAPHIC INK FOR COATED PAPER

Mix together with stirring:

Texicryl 13-011	39.501
Water	23.710
Tri-ethanolamine	3.160

Add and pass through a high speed mixer:

Reflex blue (CI42770A)	13.830

Add with stirring:

Water	19.760
Nonanol	0.039
	100.000

Texicryl 13-011 is an acrylic copolymer emulsion at 40% non-volatile content.
The solvent composition is 85% water and 15% isopropanol.
The total non-volatile content is 29%.

Initial setting of the ink will be largely due to absorption of the water by the paper which will also absorb at least some of the tri-ethanolamine. The blue pigment bleeds slightly in aqueous alkali so a very faint blue corona may surround the print. The porosity and absorbency of paper assist drying by allowing solvent to diffuse into the cellulose fibres, while its irregular surface precludes adhesion problems. Uncoated papers can cause the ink to spread on application giving a diffuse print, thus fine detail work is not practicable with such papers. In the example, nonanol is added to minimise foaming.

WHITE FLEXOGRAPHIC INK FOR PT CELLOPHANE

Mix together with stirring:	
Texicryl 13-011	33.781
Water	20.260
Ammonia solution, SG 0.880	1.350
Incorporate on a high speed mixer with:	
Rutile titanium di-oxide	16.890
Mix in with stirring:	
Isopropanol	8.440
Water	18.570
Montan wax emulsion (30% non-volatile content)	0.675
Bevaloid 581B	0.034
	100.000

After printing, the cellophane film would be heated either by passing over a heated roller, or by blowing with hot air. Alternatively, oven heating could be used, but it reduces plant flexibility.

PT cellophane is a hydrophillic surface and presents no wetting problems either to the ink, or to the dry film. It does, however, introduce drying problems because its low porosity gives a very slow rate of water absorption by the cellophane film. In the absence of absorption, drying must occur by means of solvent evaporation, and heating is applied to accelerate this process. Ammonia is used to allow complete dissociation of the salt by volatilisation of the base rendering the ink film insoluble in water. The acid form of the polymer holds water much less strongly than its ionised state, with a subsequent increase in the partial vapour pressure of the water allowing easier drying.

Note that more ammonia than the theoretical equivalent (0.9%) of the 3.16% tri-ethanolamine used in the first example. The excess is rapidly lost during drying, but it does allow some loss of base in the ink ducts and on the rollers without letting the pH drop below 8.5.

A small amount of wax is often added to aqueous inks on non-porous surfaces such as films, foils and glassine. The wax tends to migrate to the ink-air interface where it forms a non-blocking surface layer. Montan wax is used in the example and this is readily emulsified using non-ionic surfactants which are compatible with the acrylic copolymer, but emulsions of other waxes such as beeswax or carnauba could be used.

The ink has a total non-volatile content of 29% as in the first example, but the polymer non-volatile content is 15.3% (compared with 19.5%) and the solvent composition is water 80%, isopropanol 20%.

The lower polymer solids gives a lower ink viscosity which gives a thinner wet film, while the higher alcohol content in the solvent blend gives more rapid drying.

BLUE FLEXOGRAPHIC INK FOR CARTONS

Mix together with stirring:	
Vinacryl 4025	26.32
Water	39.47
Adjust pH to 7.5 with ammonia. **Incorporate on a mill with:**	
Phthalocyanine blue B	13.16
Add with stirring:	
Water	21.05
Adjust pH to 8.5 with ammonia	100.00

Vinacryl 4025 is a 50% non-volatile content emulsion of a carboxylated acrylic copolymer. The ink has a polymer non-volatile content of 15% which allows the use of water as the sole solvent without the risk of viscosity increase in the ink duct or rollers.

Any loss of ammonia during storage or printing will cause a slight drop in pH, but this will cause no significant increase in viscosity at the low solids used. The high absorbency of the board used for cartons allows rapid drying of water based inks, even when these have such low solids as this example.

RED FLEXOGRAPHIC INK FOR BOARD PRINTING

Mix together with stirring:	
Resin EP-5240	40.5
Water	10.0
Isopropanol	5.0
Nopco NXZ	1.0
Grind on a dyno mill with:	
Hoechst Red FGR	15.0
Add with stirring:	
Water	28.5
	100.0

The resin is an experimental product from Rohm & Haas which is supplied in solution form at 37% non-volatile content. The ink maker has a simple grinding and dilution operation, and is spared the trouble of neutralisation and pH adjustment.

The grinding operation is carried out at a polymer non-volatile content of 28%, which requires the presence of some isopropanol to give a reasonable viscosity. It appears that the same viscosity could be obtained by diluting with water alone, but at a lower non-volatile content, say 22%. Both this and the previous example would be suitable for four-colour printing.

OVERPRINT LACQUER FOR METAL FOIL

Mix together with stirring:	
Primal E-2212	89.9
Surfynol 104E	2.0
Neptune 1 Sp-5	1.0
Foamaster 111	0.1
Water	7.0
	100.0

The lacquer is applied to a printed foil to give both gloss and protection to the print. The Primal E-2212 is a milky solution at 40% solids with a pH of about 8.5.

This appearance is typical of a carboxylated polymer emulsion where the acid content is close to the critical value required for alkali solubility. At such levels some polymer chains will contain insufficient acid to become solubilised by alkalis and will remain in suspension. As other polymer chains in the same particle become soluble in water the remaining insoluble chains form a much smaller particle. These residual particles are partially stabilised by their ionised acid groups and give some opacity to the solution.

The other components of the lacquer are a surfactant (Surfynol) to improve wetting and a wax (Neptune 1 Sp-5) to impart slip, and reduce blocking. This type of varnish is replacing solvent based lacquers based on nitrocellulose or poly(vinyl acetate).

FLOOR POLISH APPLICATIONS

Non-slip floor polishes normally comprise a polymer emulsion mixed with a solution of a resin in dilute ammonia and a polyethylene wax. Particle flocculation occurs on drying but coalescence will only occur between similar particles because the polymer and wax are not compatible. The water soluble resin will fill the interstices between the coalesced particles but may not form a continuous phase. On ageing, polyethylene particles will tend to migrate to the air interface to reduce the surface energy of the system. In schools, offices and shops, the polish would be applied in the evening and allowed to dry overnight when some migration of the polyethylene wax would occur. During the next day the floor will be used resulting in scuffing and some marking by rubber heels. The floor will be polished mechanically restoring the gloss and removing rubber marks. The buffing operation may use a small amount of diluted polish as a lubricant. After a week or so, a fresh application of polish will be made and the cycle repeated. At intervals, the existing polish will be completely removed using special cleaners before the polish is applied. This procedure would be used on a P.V.C. floor or on a wooden surface sealed by a spar or urethane varnish. The buffed polish can have a high gloss but will not be slippery because the polyethylene wax is a hard, tough and non-deformable material. A conventional wax is either paraffinic in nature, or consists of esters of long chain alcohols and acids. These waxes are slippery because the weak intermolecular forces between their molecules give low resistance to stress.

Polyethylene emulsions can be prepared by melting the oxidised polyethylene with about 10% of its weight of oleic acid and adding the melt to a vigorously stirred alkaline solution at 75−90°C. The final pH would be about 9.0 and the particle size 0.1 micron.

In order to avoid destabilisation of the wax emulsion the pH of the polymer emulsion is adjusted to 9−9.5 before mixing. One of the side effects of ionising the acid groups in the emulsion polymer is to confer stability to freezing.

This is termed the freeze-thaw stability and is normally expressed as the number of cycles of freezing, and thawing, an emulsion will undergo before coagulating.

STYRENE-SHELLAC POLISH

Polystyrene emulsion	72.50
Add slowly with stirring:	
Tri-butoxyethyl phosphate	2.42
Mix in:	
Polyethylene wax emulsion (30%)	24.17
FC128 (1% solution)	0.91
	100.00

The styrene shellac latex polymer described on page 260, is suitable for this polish. Even with the plasticiser present, the styrene homopolymer is too brittle to form a film and the polish consists essentially of a shellac film, containing particles of styrene polymer with particles of oxidised polyethylene concentrated on the surface. The hard polystyrene particles give the film hardness and the shellac film induces toughness. The polish is rather dark and further darkens on ageing which also results in insolubilisation of the shellac due to condensation type polymerisation.

ACRYLIC FLOOR POLISH

Mix together:	
Acrylic emulsion (36% non-volatile content)	44.34
Waterez 1582 solution (50%)	18.87
Epolene E10 emulsion (30%)	33.02
Add slowly with stirring:	
Tri-butoxyethyl phosphate	2.83
Mix in:	
FC128 solution (1%)	0.94
	100.00

A typical copolymer would comprise 40% methyl methacrylate, 33% methyl acrylate, 25% styrene and 2% acrylic acid. The emulsion would be stabilised entirely by anionic surfactant and have a particle size of about 0.09 micron, and the polymer has a Tg of 36°C. Most of the polymers used in floor polishes have a Tg above ambient.

Tri-butoxyethyl phosphate is widely used as a plasticiser in floor polishes. There seems little justification on technical grounds for its widespread use. The use of a fluorocarbon surfactant is justified. FC128 is the sodium salt of perfluoroheptyl methyl sulphate $C_7F_{15}.CH_2.O.SO_3Na$ which gives a surface tension of about 18 dynes per cm, which is very much lower than that given by conventional surfactants. While other surfactants will concentrate on oil-water interfaces to give a system with the lowest surface energy, perfluoro surfactants have no affinity with polymers or oils in which the hydrophobic group is of a hydrocarbon nature. Consequently fluorocarbon surfactants remain in aqueous solution and are not absorbed by polymer particles. This polish is readily polished by buffing and easily removed by alkaline solutions, however long it has aged.

The resin solution of the example on the next page, is a low molecular weight (2140) styrene acrylic acid copolymer adjusted to a non-volatile content of 20% and pH 9.1.

The zinc is stabilised as an amine complex while an excess of ammonia is present in solution. On drying, the ammonia evaporates and the zinc forms a salt with the acid groups in the polymer backbone. Zinc is divalent and reacts with two carboxylic acid groups.

The zinc acid bond is largely covalent in the absence of water and causes extensive cross-linking. The quantity of zinc added will react with about 55% of the acid groups in the polymer. Cross-linking gives an increase in film strength and particularly its toughness. Another effect of cross-linking is to reduce the amount of solvent which the film can absorb. The practical result of this property is to reduce the sensitivity of the film to alkalis.

Many floor cleaners are based on sodium metasilicate as the alkaline component. Neither in the ionised nor molecular form will this alkali penetrate a cross-linked film and solubilise the polymer by ionising its acid groups. Metallised finishes of this type

METAL-COMPLEXED ACRYLIC POLISH

Mix together:	
Resin solution (20%)	21.2
Polymer emulsion (35%)	18.4
Surfactant (nonyl phenol + 9 EO)	1.0
Water	54.4
Propylene glycol	2.0
Ammonia solution (16.5 N)	2.4
Add slowly with stirring:	
Tri-butoxyethyl phosphate	0.4
Add and stir until dissolved:	
Zinc oxide	0.2
	100.0

can be recoated without the blushing found with other polishes and this leads to better adhesion between successive coats. The polish can be removed by a cleaner using ammonia or an organic amine as the alkaline component, and such a base can complex with the zinc atoms and is sufficiently soluble in the polymer phase to diffuse into the polish film.

Zirconium salts also cross-link polymer films by reaction with acid groups and can be used instead of zinc in this application.

These three examples show the range of polymer types used in floor polishes; the two polymers with acid groups contain two and five percent respectively of acrylic acid which is insufficient to solubilise the polymer when made alkaline. It also shows three types of alkali soluble resins which act as levellers in polish formulations. Other resins which could be used include alkyd or polyester resins with relatively high acid values. There is some evidence that the addition of a solvent such as white spirit to the polish will assist in removing rubber marks, but the practise has not been widely adopted.

PAPER COATING

The coating is essentially a dispersion of china clay or sometimes satin white in a latex at a high pigment volume content. It is applied to the paper by rollers and excess coating is removed by a blade or, less commonly, by an air knife. The coating reduces the porosity of the paper, essentially by replacing the paper surface by a coating surface. This permits better detail to be obtained during printing. Coating also improves the appearance of the paper.

GRAVURE QUALITY PAPER
FOR BLADE COATING

Dinkie A clay	53.927
Dispex N40	0.108
Oxidised starch solution (10%)	16.180
Intex 177	10.780
Calcium stearate	0.135
Water	18.870
	100.000

These components would be blended in a Silverson type mixer. The non-volatile content would be 63% and the binder:pigment ratio 13%. Note the use of binder:pigment ratio rather than the pigment volume concentration favoured in paints.

Intex 177 contains a carboxylated styrene-butadiene copolymer at 50% non-volatile content. The emulsion has a particle size of about 0.15 micron. The small particle size gives improved gloss at low p.v.c., and better binding power at high pigment loading.

Styrene-butadiene latices produce stable colour mixes and good pigment binding. The coating has high resistance to water whether measured as wet rub or wet pick tests and gives good gloss on calendering.

BLADE COATING FOR FLEXOGRAPHIC PAPER

Dinkie A clay	53.055
Dispex N40	0.265
Pluronic 61	0.530
Oxidised starch solution (10%)	15.910
National 125 – 1104	10.610
Aerosol 22 (25% non-volatile content)	1.060
Water	18.570
	100.000

Blended on a Silverson mixer.

Non-volatile content 63%; binder:pigment ratio 13%

National 125–1104 is a polyvinyl acetate emulsion at 50% non-volatile content of particle size 0.17 micron and pH 7.5.

Polyvinyl acetate emulsions give coating mixtures which thicken on standing, and to minimise this effect extra dispersants and surfactants are added. In this instance the Aerosol is an anionic surfactant and the Pluronic is nonionic. Pluronic 61 has a low H.L.P. (Hydrophile-lyophile balance) and also acts as an anti-foaming agent. The more hydrophillic nature of polyvinyl acetate compared with styrene-butadiene, gives poorer wet rub and wet pick than the previous example.

The advantage of the polyvinyl acetate is its printability due to its high surface energy. The critical surface tension of polyvinyl acetate is about 40 dynes per cm which is higher than the surface tension of most aqueous printing inks which range from 35−39.

The critical surface tension of styrene-butadiene copolymers is about 34−35 dynes per cm and they are not always wetted by flexographic inks. A surface is only wetted by a liquid whose surface tension is lower than the critical surface tension of the surface.

CHROMO PAPER BY AIR-KNIFE COATING

SPS clay	23.010
Calgon S	0.066
Satin white	9.860
Casein solution, ammonia (10%)	32.874
Intex 177	7.890
Water	26.300
	100.000

Blended by Silverson mixer.

Non-volatile content 40%; binder:pigment ratio 22%

An air knife cannot exert the force on the coating that a fixed steel blade can. Air knife coatings have much lower viscosity than blade coatings. This is achieved by dilution with water. The coating will be thicker than that normally achieved by blade coating. A heavy coating gives an art paper often termed a 'chromo'.

The use of casein provides a solution polymer which would normally produce a continuous film. SBR is not compatible with casein so a casein film containing both pigment and polymer particles would be expected. There is insufficient casein available to form such a continuous film and it will tend to concentrate on the pigment surface forming some bridges between pigment particles. The styrene-butadiene polymer will form a continuous film surrounding pigment particles and the casein bound pigment agglomerates. The properties will be largely controlled by the polymer which is readily wetted by the lithographic inks which would be used for this grade of paper.

INDEX TO TRADE NAMES

Acrysol	Rohm & Haas Company
Aerosol	American Cyanamide Company
Araldite	Ciba Geigy
Asbestine	
Balab	Balab Incorporated
Bevaloid	Richard Hodgson & Sons Limited
Carbitol	Union Carbide Chemical Company
Casamid	Thomas Swan & Company Limited
Cellosize	Union Carbide Chemical Company
Cellosolve	Union Carbide Chemical Company
Calgon	Albright & Wilson Limited
Dispex	Allied Colloids Limited
Dynomin	Dyno Industrier
Dynotal	Dyno Industrier
Epikote	Shell Chemical Limited
Epilink	AZKO Chemie
Epolene	Eastman Kodak Company
Foamaster	Diamond Shamrock Company
Genklene	ICI
Hypalon	Du Pont
Intex	International Synthetic Rubber Company
Kronos	Kronos Titanium Pigments Limited
Manoxol	Hardman & Holden Limited
Microdol	Norwegian Talc (U.K.) Limited
Microtalc	Norwegian Talc (U.K.) Limited
Mowital	Hoescht AG
Natrosol	Hercules Incorporated
Nonanol	ICI
Nopco	Nopco Hess Limited
National	National Adhesives & Resins Limited
Neptune	Shamrock Chemicals
Pluronic	Wyandotte Corporation
Polidene	Scott Bader & Company Limited
Rhoplex	Rohm & Haas Company
Shellsol	Shell Chemicals Limited
Surfynol	American Cyanamide Company
Syloid	Bayer
Tamol	Rohm & Haas Company
Texicryl	Scott Bader & Company Limited
Tiona	Laporte Industries Limited
Tioxide	BTP Tioxide Limited
Triton	Rohm & Haas Company
Veova	Shell Chemicals Limited
Versamid	Cray Valley Products Limited
Vinacryl	Vinyl Products
Vinamul	Vinyl Products
Vinylite	Bakelite Limited
Wolfamid	NL Victor Wolf Limited
Wolfkur	NL Victor Wolf Limited

INDEX